Self-Tuning Control for
Two-Dimensional Processes

UMIST CONTROL SYSTEMS CENTRE SERIES
Series Editors: **Dr M. B. Zarrop** *and* **Professor P. E. Wellstead**
   UMIST, UK

   1. Change Detection and Input Design in Dynamical Systems
      Feza Kerestecioğlu

   2. Self-Tuning Control for Two-Dimensional Processes
      W. P. Heath

Self-Tuning Control for Two-Dimensional Processes

W. P. Heath
Control Systems Centre, UMIST, UK

RESEARCH STUDIES PRESS LTD.
Taunton, Somerset, England

JOHN WILEY & SONS INC.
New York · Chichester · Toronto · Brisbane · Singapore

RESEARCH STUDIES PRESS LTD.
24 Belvedere Road, Taunton, Somerset, England TA1 1HD

Copyright © 1994 by Research Studies Press Ltd.

All rights reserved.

No part of this book may be reproduced by any means,
nor transmitted, nor translated into a machine language
without the written permission of the publisher.

Marketing and Distribution:

Australia and New Zealand:
Jacaranda Wiley Ltd.
GPO Box 859, Brisbane, Queensland 4001, Australia

Canada:
JOHN WILEY & SONS CANADA LIMITED
22 Worcester Road, Rexdale, Ontario, Canada

Europe, Africa, Middle East and Japan:
JOHN WILEY & SONS LIMITED
Baffins Lane, Chichester, West Sussex, England

North and South America:
JOHN WILEY & SONS INC.
605 Third Avenue, New York, NY 10158, USA

South East Asia:
JOHN WILEY & SONS (SEA) PTE LTD.
37 Jalan Pemimpin 05-04
Block B Union Industrial Building, Singapore 2057

Library of Congress Cataloging-in-Publication Data

Heath, W. P. (William Paul), 1966–
    Self-tuning control for two-dimensional processes / W.P. Heath.
        p. cm. — (UMIST Control Systems Centre series ; 2)
    Originally presented as the author's thesis (doctoral—University
of Manchester), 1992.
    Includes bibliographical references and index.
    ISBN 0-86380-166-8 (Research Studies Press, LTD.). — ISBN
0-471-95135-8 (John Wiley & Sons Inc., New York)
    1. Self-tuning controllers. 2. Process control. I. Title.
II. Series.
TJ217.H43   1994
629.8'312—dc20   94-7741
                                                   CIP

British Library Cataloguing in Publication Data

A catalogue record for this book
is available from the British Library.

ISBN 0 86380 166 8 (Research Studies Press Ltd.)
ISBN 0 471 95135 8 (John Wiley & Sons Inc.)

Printed in Great Britain by SRP Ltd., Exeter

Editorial Foreword

For over 25 years the Control Systems Centre at UMIST has been at the forefront of research in the key areas of control, signal processing and - more recently - information technology. It is the intention of this book series to make more widely and more speedily available the best of the current contributions in these areas emerging from the Centre and, where appropriate, from collaborating individuals or organisations. Each volume will aim to present new results and an up-to-date survey of a particular area of current or potential importance to researchers and the wider engineering community.

Martin Zarrop
Peter Wellstead

Control Systems Centre
Department of Electrical
Engineering and Electronics
University of Manchester
Institute of Science and Technology

Preface

In this monograph we develop least squares optimal prediction, minimum variance control and generalised minimum variance control algorithms for a two-dimensional CARMA process. Each algorithm involves the algebraic solution of a two-dimensional Diophantine equation whose solution polynomials are in general of infinite degree. However, these solution polynomials may also be expressed as rational transfer functions of finite degree and hence the predictor and controller algorithms may be implemented in closed form. The algorithms must be modified for any practical implementation to take into account the edges of the data field. In this case the process may be analysed using multivariable theory, and hence linkages between multivariable representations and two-dimensional systems can be set up.

The algorithms are suitable for setting in self-tuning mode by combining them with recursive parameter estimation algorithms. We consider the problem of introducing forgetting factors to parameter estimation algorithms. By placing previous estimator algorithms on a sound theoretical basis we show that conventional methods for forgetting are appropriate only for very specific data structures; thus we introduce a more general algorithm that forgets in two dimensions. We also consider the problems of setpoint tracking and offset handling for the self-tuning controllers. We show that if the two-dimensional process can be modelled as incremental then the solution is particularly straightforward.

Throughout we illustrate the results with simulations.

Acknowledgements

First and foremost I owe an enormous debt of gratitude to Prof. Peter Wellstead for all his help and advice. His constant encouragement and tireless efforts on my behalf have been an inspiration and a blessing.

Thanks must also go to Dr Martin Zarrop of the Control Systems Centre and Dr Ruda Kulhavy (on sabbatical from the Czechoslovak Academy of Sciences, Prague) who have both shown a great interest in the project and an invaluable attention to detail. I would also like to thank the many other students and staff at the Control Systems Centre for their help, and in particular Xabier Troyas and Allan Kjaer for many useful and lively discussions.

I would like to thank Dr Gaynor Taylor of Hull University and Dr Clive Pugh of Loughborough University for their guidance through some of the algebra of two-dimensional systems. Similarly I would like to thank Dr Howard Kaufman of the Rensselaer Polytechnic Institute, New York, for some useful discussions concerning two-dimensional estimation.

Dr Mike Waller of Miami University, Ohio, and Peter Herdman of Wiggins Teape Research and Development Ltd introduced me to the basis weight control problem in papermaking. Similarly Dr Roger Edgar of Infrared Engineering Ltd introduced me to the plastic film extrusion process. Thanks go to them all.

This work was originally submitted as a thesis [Heath 1992] to the University of Manchester for the degree of Doctor of Philosophy, January 1992. It was sponsored by the SERC on research grant GR/F 01338 and also in part by NATO grant 0124/87. The Royal Society enabled me to spend a week at the Institute of Information Theory and Automation, Czechoslovak Academy of Sciences, Prague.

Contents

List of theorems	xi
List of algorithms	xii
List of tables	xiii
List of figures	xiv
Selective glossary of abbreviations and symbols	xxi

1 Introduction — 1
   1.1 Background — 1
   1.2 Applications — 4
   1.3 Structure of the monograph — 6
   1.4 Contribution — 7

2 Prediction and control over the semi-infinite plane — 9
   2.1 The two-dimensional control problem — 9
   2.2 Some results from two-dimensional systems theory — 11
      2.2.1 Transforming from quarter-plane to non-symmetric half-plane causal processes — 12
      2.2.2 Zeros and coprimeness — 13
      2.2.3 Stability — 14
      2.2.4 Autocovariance of an ARMA process — 15
      2.2.5 The solution of two-dimensional Diophantine equations — 17
   2.3 Pole-assignment control — 19
   2.4 Prediction — 21
      2.4.1 The two-dimensional prediction problem — 21
      2.4.2 A closed form for the least squares predictor — 25
      2.4.3 An algebraic interpretation — 26
   2.5 Minimum variance and generalised minimum variance control — 28
      2.5.1 Minimum variance control — 28

| 2.5.2 Generalised minimum variance control | 30 |

3 Prediction and control over the plane of finite width — 32
3.1 Prediction and control for case with edges — 32
3.1.1 The process — 32
3.1.2 Prediction and minimum variance control — 34
3.1.3 The relationship with the case with no edges — 37
3.1.4 Generalised minimum variance control — 39
3.2 Simulations — 40
3.3 The Multivariable Connection — 43
3.3.1 The process — 43
3.3.2 Prediction — 47
3.3.3 Minimum variance control — 50
3.3.4 Generalised minimum variance control — 51
3.3.5 The link with the case without edges — 54
3.3.6 Reprise — 56
3.3.7 Other models for behaviour at the edges — 59
3.4 Other causality structures — 61
3.4.1 The process — 61
3.4.2 Model responses to step inputs — 63
3.4.3 Control—an informal approach — 70
3.4.4 A simulation example — 74
3.4.5 Control—a more formal approach — 75

4 Parameter estimation — 83
4.1 Least squares estimation in two dimensions — 83
4.1.1 AR processes — 83
4.1.2 The extension to ARMA data — 87
4.2 Forgetting strategies — 88
4.3 Two-dimensional forgetting with $0 < \lambda < 1$ — 92
4.3.1 An 'Attasi's model' form for two-dimensional forgetting — 92
4.3.2 A 'Roesser's model' form for two-dimensional forgetting — 96
4.4 Two-dimensional forgetting in factored form — 102
4.5 Row and column forgetting — 110
4.5.1 Row forgetting — 110

4.5.2 Column forgetting—the basic algorithm	112
4.5.3 Column forgetting—modifications	113
4.6 Simulations	116

5 Self-tuning control — 136

    5.1 Self-tuning GMV in two dimensions — 137
        5.1.1 An explicit self-tuning controller — 137
        5.1.2 Implicit self-tuning control — 144
        5.1.3 Self-tuning control for different edge conditions — 146
    5.2 Setpoint tracking — 149
        5.2.1 Tracking without edges — 149
        5.2.2 Tracking column by column — 151
        5.2.3 Integral control — 153
    5.3 Offset handling — 158
        5.3.1 The incremental model — 159
        5.3.2 The static model — 162

Conclusion — 170

Appendix 1. Proof of the theorems — 174

Appendix 2. Algorithm 2.4.1 — 198

Appendix 3. Multivariable self-tuning control — 201

    A3.1 The process — 201
    A3.2 Prediction — 201
    A3.3 Minimum variance control — 202
    A3.4 Generalised minimum variance control — 202

Appendix 4. Simulation results for §4.6 — 204

References — 229

Index — 235

List of theorems

Theorem 2.2.1 *(Preservation of covariances under co-ordinate transformation).*   15
    Theorem 2.2.2 *Hilbert's Nullstellensatz.*   17
    Theorem 2.2.3 *Bezout's Theorem.*   18
Theorem 2.4.1 *(Existence of finite-order polynomials Φ^{k_1,k_2} and Γ^{k_1,k_2}).*   25
    Theorem 2.4.2 *(Least squares prediction).*   25
    Theorem 2.4.3 *(Prediction-error variance).*   26
Theorem 2.5.1 *(Minimum variance control).*   29
    Theorem 2.5.2 *(Generalised minimum variance control).*   30

Theorem 3.1.1 *(Minimum variance control on a plane of finite width).*   36
    Theorem 3.1.2 *(Relation between F^{k_1,k_2} and F_{m,k_1,k_2}).*   38
    Theorem 3.1.3 *(Generalised minimum variance control on a plane of finite width).*   39
Theorem 3.3.1 *(Minimum variance control applied successively).*   51
    Theorem 3.3.2 *(Relation between two-dimensional and multivariable minimum variance control).*   56

Theorem 4.4.1 *(Rank-one update of orthogonal decomposition).*   103
    Theorem 4.4.2 *(Generalisation of Theorem 4.4.1 to updates of any rank).*   104
    Theorem 4.4.3 *(Generalisation of Theorem 4.4.2 to fast Givens transformations).*   108

Theorem 5.2.1 *(Use of integral control when $B(w^{-1}, z^{-1})$ is symmetric).*   157

List of algorithms

Algorithm 2.3.1 *Two-dimensional pole-assignment control.*   20

Algorithm 2.4.1 *Calculation of* $\Phi^{k_1,k_2}(w^{-1},z^{-1})$.   25

(199

Algorithm 4.1.1 *2D-RLS.*   86

Algorithm 4.3.1 *Basic two-dimensional forgetting factors.*   95

    Algorithm 4.3.2 *Two-dimensional forgetting without subtraction: first form.*   100

    Algorithm 4.3.3 *Two-dimensional forgetting without subtraction: second form.*   101

Algorithm 4.4.1 *Factored form of 4.3.2 using Householder transformations.*   106

    Algorithm 4.4.2 *Factored form of 4.3.2 using fast Givens rotations.*   108

Algorithm 4.5.1 *Two-dimensional forgetting when $\lambda = 0$.*   112

    Algorithm 4.5.2 *Separate estimators over separate columns.*   114

Algorithm 5.1.1 *Explicit two-dimensional self-tuning generalised minimum variance control.*   137

Algorithm 5.1.2 *Implicit two-dimensional self-tuning generalised minimum variance control.*   145

List of tables

Table 4.6.1 *Chosen values of forgetting factors for forgetting algorithms.* 117

Table 4.6.2 *Figures for the meshes.* 126

Table 4.6.3 *Figures for the cross-sections.* 127

List of figures

Fig 1.2.1 *Representation of the generic-web manufacturing process.*   4

    Fig 1.2.2 *Representation of the papermaking process.*   5

Fig 2.1.1 *Global and local supports for a two-dimensional NSHP process.*   10

    Fig 2.1.2 *The raster scan.*   11

Fig 2.3.1 *Zeros of the polynomials $A(w^{-1}, z^{-1})$, $z^{-1}B(w^{-1}, z^{-1})$ and $T(w^{-1}, z^{-1})$.*   22

Fig 2.4.1 *Pixel to be predicted.*   23

    Fig 2.4.2 *Prediction 'along the row'.*   24

    Fig 2.4.3 *Heuristic interpretation of the significance of $A(w^{-1}, 0)$.*   28

Fig 3.1.1 *Global support with edges.*   33

    Fig 3.1.2 *The truncation of local supports.*   33

    Fig 3.1.3 *The finite region between the predicted value and the known data.*   35

    Fig 3.1.4 *Modified local support for $A(w^{-1}, z^{-1})$.*   37

    Fig 3.1.5 *Finite support for $F^{k_1, k_2}(w^{-1}, z^{-1})$ when $A(w^{-1}, 0) = 1$.*   38

Fig 3.2.1 *Cumulative squared output for minimum variance control with $A_i = A_1$ and $B_i = B_1$.*   42

    Fig 3.2.2 *The difference in outputs between the two minimum variance controllers with $A_i = A_1$ and $B_i = B_1$.*   42

    Fig 3.2.3 *Cumulative squared output for minimum variance control with $A_i = A_2$ and $B_i = B_1$*   44

    Fig 3.2.4 *The difference in outputs between the two minimum variance controllers with $A_i = A_2$ and $B_i = B_1$.*   44

    Fig 3.2.5 *Cumulative squared output for generalised minimum variance control with $A_i = A_2$ and $B_i = B_2$.*   45

    Fig 3.2.6 *The difference in outputs between the two generalised minimum variance controllers with $A_i = A_2$ and $B_i = B_2$.*   45

Fig 3.3.1 *Two-dimensional predictor.*   48

Fig 3.3.2 *Multivariable predictor.* 48

Fig 3.3.3 *The modified multivariable predictor.* 49

Fig 3.4.1 *NSHP (non-symmetric half-plane) local support.* 61

Fig 3.4.2 *SHP (symmetric half-plane) local support.* 62

Fig 3.4.3 *SWP (symmetric wedge plane) local support.* 63

Fig 3.4.4 *Local supports.* 64

Fig 3.4.5 *Step responses with $A(w^{-1}, z^{-1})$ SHP-causal.* 65

Fig 3.4.6 *Initial responses to step input with $A(w^{-1}, z^{-1})$ SHP-causal.* 66

Fig 3.4.7 *Step responses with $A(w^{-1}, z^{-1})$ NSHP-causal.* 66

Fig 3.4.8 *Initial responses to step input with $A(w^{-1}, z^{-1})$ NSHP-causal.* 67

Fig 3.4.9 *Step responses with $A(w^{-1}, z^{-1})$ SWP-causal.* 67

Fig 3.4.10 *Initial responses to step input with $A(w^{-1}, z^{-1})$ SWP-causal.* 68

Fig 3.4.11 *A single actuator (screw) change causing the lip profile to change over a wide area.* 69

Fig 3.4.12 *Relation of outputs to inputs at the righthand edge for the process modelled by Equations 3.4.1 and 3.4.2.* 70

Fig 3.4.13 *Pictorial representation of Equation 3.4.3.* 73

Fig 3.4.14 *Pictorial representation of Equations 3.4.5.* 73

Fig 3.4.15 *Pictorial representation of Equation 3.4.6 formed by substituting from Equations 3.4.4 successively into Equation 3.4.3.* 74

Fig 3.4.16 *Closed-loop output when $B(w^{-1}, z^{-1})$ has SHP support.* 76

Fig 3.4.17 *Input when $B(w^{-1}, z^{-1})$ has SHP support.* 76

Fig 3.4.18 *Support for $F_1(w^{-1}, z^{-1})$.* 78

Fig 3.4.19 *Support for $F_2(w^{-1}, z^{-1})$.* 78

Fig 3.4.20 *Closed-loop output when $B(w^{-1}, z^{-1})$ has SHP support using the 'optimal' controller.* 81

Fig 3.4.21 *Input when $B(w^{-1}, z^{-1})$ has SHP support using the 'optimal' controller.* 81

Fig 4.1.1 *Regions where estimates of the noise $e(m,n)$ may be required but are not explicitly calculated in 2D-RELS or 2D-AML.* 88

Fig 4.2.1 *Weighting of pixel information in previous forgetting strategies.* 89

Fig 4.2.2 *Weighting of pixel information in the new forgetting strategy.* 90

Fig 4.2.3 *The 'support regions' for $Q_l(m,n)$ and $Q_r(m,n)$.* 91

Fig 4.3.1 *The update of $Q_l(m,n)$.* 94
Fig 4.3.2 *The update of $Q_r(m,n)$.* 94
Fig 4.3.3 *The updating of $Q_l(m,n)$ with the scan.* 96
Fig 4.3.4 *The updating of a row of $Q_r(m,n)$'s.* 97
Fig 4.3.5 $Q_l^h(m,n)$, $Q_r^h(m,n)$ and $Q^v(m,n-1)$. 100
Fig 4.5.1 *Regions for Algorithm 4.5.2.* 114
Fig 4.6.1 *Cases (i) and (ii)—horizontal and diagonal changes respectively.* 118
Fig 4.6.2 *Cases (iii) and (iv)—vertical and circular changes respectively.* 118
Fig 4.6.3 *'Mesh' of $\hat{a}$ for case (i) using general two-dimensional forgetting.* 119
Fig 4.6.4 *'Mesh' of $\hat{c}$ for case (i) using general two-dimensional forgetting.* 119
Fig 4.6.5 *Vertical cross-section of $\hat{a}$ and $\hat{c}$ for case (i) using general two-dimensional forgetting.* 120
Fig 4.6.6 *'Mesh' of $\hat{a}$ for case (iii) using general two-dimensional forgetting.* 120
Fig 4.6.7 *'Mesh' of $\hat{c}$ for case (iii) using general two-dimensional forgetting.* 121
Fig 4.6.8 *Horizontal cross-section of $\hat{a}$ and $\hat{c}$ for case (iii) using general two-dimensional forgetting.* 121
Fig 4.6.9 *'Mesh' of $\hat{a}$ for case (i) using Wagner's algorithm.* 122
Fig 4.6.10 *'Mesh' of $\hat{c}$ for case (i) using Wagner's algorithm.* 122
Fig 4.6.11 *Vertical cross-section of $\hat{a}$ and $\hat{c}$ for case (i) using Wagner's algorithm.* 123
Fig 4.6.12 *'Mesh' of $\hat{a}$ for case (iii) using Wagner's algorithm.* 123
Fig 4.6.13 *'Mesh' of $\hat{c}$ for case (iii) using Wagner's algorithm.* 124
Fig 4.6.14 *Horizontal cross-section of $\hat{a}$ and $\hat{c}$ for case (iii) using Wagner's algorithm.* 124
Fig 4.6.15 *Edge changes.* 128
Fig 4.6.16 *Horizontal cross-section of $\hat{a}$ and $\hat{c}$ for case (v) using general two-dimensional forgetting.* 128
Fig 4.6.17 *Horizontal cross-section of $\hat{a}$ and $\hat{c}$ for case (v) using column forgetting.* 129
Fig 4.6.18 *Horizontal cross-section of $\hat{a}$ and $\hat{c}$ for case (v) using column forgetting at the edges.* 129
Fig 4.6.19 *Vertical cross-sections of $\hat{a}$ parameters for case (i) with eight parameters using general two-dimensional forgetting with λ and ν set to 0.77.* 130

Fig 4.6.20 *Vertical cross-sections of $\hat{c}$ parameters for case (i) with eight parameters using general two-dimensional forgetting with λ and ν set to 0.77.*   130

Fig 4.6.21 *Horizontal cross-sections of $\hat{a}$ parameters for case (iii) with eight parameters using general two-dimensional forgetting with λ and ν set to 0.77.*   131

Fig 4.6.22 *Horizontal cross-sections of $\hat{c}$ parameters for case (iii) with eight parameters using general two-dimensional forgetting with λ and ν set to 0.77.*   131

Fig 4.6.23 *Vertical cross-sections of $\hat{a}$ parameters for case (i) with eight parameters using Wagner's algorithm with λ set to 0.98.*   132

Fig 4.6.24 *Vertical cross-sections of $\hat{c}$ parameters for case (i) with eight parameters using Wagner's algorithm with λ set to 0.98.*   132

Fig 4.6.25 *Horizontal cross-sections of $\hat{a}$ parameters for case (iii) with eight parameters using Wagner's algorithm with λ set to 0.98.*   133

Fig 4.6.26 *Horizontal cross-sections of $\hat{c}$ parameters for case (iii) with eight parameters using Wagner's algorithm with λ set to 0.98.*   133

Fig 4.6.27 *Vertical cross-sections of $\hat{a}$ parameters for case (i) with eight parameters using general two-dimensional forgetting with λ and ν set to 0.95.*   134

Fig 4.6.28 *Vertical cross-sections of $\hat{c}$ parameters for case (i) with eight parameters using general two-dimensional forgetting with λ and ν set to 0.95.*   134

Fig 4.6.29 *Horizontal cross-sections of $\hat{a}$ parameters for case (iii) with eight parameters using general two-dimensional forgetting with λ and ν set to 0.95.*   135

Fig 4.6.30 *Horizontal cross-sections of $\hat{c}$ parameters for case (iii) with eight parameters using general two-dimensional forgetting with λ and ν set to 0.95.*   135

Fig 5.1.1 *Closed-loop output for the basic self-tuning two-dimensional generalised minimum variance controller.*   140

Fig 5.1.2 *Cumulative output squared for the generalised minimum variance controller.*   140

Fig 5.1.3 *Parameter estimates for $A(w^{-1}, z^{-1})$.* 141

Fig 5.1.4 *Parameter estimates for $B(w^{-1}, z^{-1})$.* 141

Fig 5.1.5 *Parameter estimates for $C(w^{-1}, z^{-1})$.* 142

Fig 5.1.6 *Parameter estimates for the controller polynomial $X_{m,0,\nu}(w^{-1}, z^{-1})$ evaluated away from the edges.* 142

Fig 5.1.7 *Parameter estimates for the controller polynomial $Y_{m,0,\nu}(w^{-1}, z^{-1})$ evaluated away from the edges.* 143

Fig 5.1.8 *Parameter estimates for the controller polynomial $Z_{m,0,\nu}(w^{-1}, z^{-1})$ evaluated away from the edges.* 143

Fig 5.1.9 *Cumulative output squared comparing explicit and implicit controllers for the process given by Equations 5.1.4 to 5.1.7.* 147

Fig 5.1.10 *Closed-loop output for the implicit and explicit controllers on the process given by Equations 5.1.8 to 5.1.11.* 147

Fig 5.1.11 *Cumulative output squared for three controllers with a process whose behaviour changes at the edges.* 149

Fig 5.2.1 *Output with setpoint 100 with scalar weighting.* 151

Fig 5.2.2 *Output with setpoint 100 with each column weighted separately.* 153

Fig 5.2.3 *Input with setpoint 100 with each column weighted separately.* 154

Fig 5.2.4 *The weightings for each column.* 154

Fig 5.2.5 *Output with setpoint 100 with integral control.* 155

Fig 5.2.6 *Input with setpoint 100 with integral control.* 156

Fig 5.3.1 *Output with setpoint 100 for the incremental model, parameters known.* 162

Fig 5.3.2 *Input with setpoint 100 for the incremental model, parameters known.* 163

Fig 5.3.3 *Output with setpoint 100 for the incremental model, parameters estimated.* 164

Fig 5.3.4 *Output with setpoint 100 for the static model, parameters known.* 167

Fig 5.3.5 *Input with setpoint 100 for the static model, parameters known.* 168

Fig 5.3.6 *Output with setpoint 100 for the static model, parameters estimated.* 168

Fig 5.3.7 *Output with setpoint 100 for the static model, parameters estimated for reduced order model.* 169

Fig 5.3.8 *Offset values.* 169

Fig A1.1 *Terms in $P(k)$.* 186

Fig A1.2 *The delay in the process.* 187

Fig A4.1 *'Mesh' of $\hat{a}$ for case (ii) using general two-dimensional forgetting.* 204
Fig A4.2 *'Mesh' of $\hat{c}$ for case (ii) using general two-dimensional forgetting.* 204
Fig A4.3 *Vertical cross-section of $\hat{a}$ and $\hat{c}$ for case (ii) using general two-dimensional forgetting.* 205
Fig A4.4 *Horizontal cross-section of $\hat{a}$ and $\hat{c}$ for case (ii) using general two-dimensional forgetting.* 205
Fig A4.5 *'Mesh' of $\hat{a}$ for case (iv) using general two-dimensional forgetting.* 206
Fig A4.6 *'Mesh' of $\hat{c}$ for case (iv) using general two-dimensional forgetting.* 206
Fig A4.7 *Vertical cross-section of $\hat{a}$ and $\hat{c}$ for case (iv) using general two-dimensional forgetting.* 207
Fig A4.8 *Horizontal cross-section of $\hat{a}$ and $\hat{c}$ for case (iv) using general two-dimensional forgetting.* 207
Fig A4.9 *'Mesh' of $\hat{a}$ for case (ii) using Wagner's algorithm.* 208
Fig A4.10 *'Mesh' of $\hat{c}$ for case (ii) using Wagner's algorithm.* 208
Fig A4.11 *Vertical cross-section of $\hat{a}$ and $\hat{c}$ for case (ii) using Wagner's algorithm.* 209
Fig A4.12 *Horizontal cross-section of $\hat{a}$ and $\hat{c}$ for case (ii) using Wagner's algorithm.* 209
Fig A4.13 *'Mesh' of $\hat{a}$ for case (iv) using Wagner's algorithm.* 210
Fig A4.14 *'Mesh' of $\hat{c}$ for case (iv) using Wagner's algorithm.* 210
Fig A4.15 *Vertical cross-section of $\hat{a}$ and $\hat{c}$ for case (iv) using Wagner's algorithm.* 211
Fig A4.16 *Horizontal cross-section of $\hat{a}$ and $\hat{c}$ for case (iv) using Wagner's algorithm.* 211
Fig A4.17 *'Mesh' of $\hat{a}$ for case (i) using column forgetting.* 212
Fig A4.18 *'Mesh' of $\hat{c}$ for case (i) using column forgetting.* 212
Fig A4.19 *Vertical cross-section of $\hat{a}$ and $\hat{c}$ for case (i) using column forgetting.* 213
Fig A4.20 *'Mesh' of $\hat{a}$ for case (ii) using column forgetting.* 213
Fig A4.21 *'Mesh' of $\hat{c}$ for case (ii) using column forgetting.* 214
Fig A4.22 *Vertical cross-section of $\hat{a}$ and $\hat{c}$ for case (ii) using column forgetting.* 214
Fig A4.23 *Horizontal cross-section of $\hat{a}$ and $\hat{c}$ for case (ii) using column forgetting.* 215
Fig A4.24 *'Mesh' of $\hat{a}$ for case (iii) using column forgetting.* 215
Fig A4.25 *'Mesh' of $\hat{c}$ for case (iii) using column forgetting.* 216

Fig A4.26 *Horizontal cross-section of â and ĉ for case (iii) using column forgetting.*	216
Fig A4.27 *'Mesh' of â for case (iv) using column forgetting.*	217
Fig A4.28 *'Mesh' of ĉ for case (iv) using column forgetting.*	217
Fig A4.29 *Vertical cross-section of â and ĉ for case (iv) using column forgetting.*	218
Fig A4.30 *Horizontal cross-section of â and ĉ for case (iv) using column forgetting.*	218
Fig A4.31 *'Mesh' of â for case (i) using row forgetting.*	219
Fig A4.32 *'Mesh' of ĉ for case (i) using row forgetting.*	219
Fig A4.33 *Vertical cross-section of â and ĉ for case (i) using row forgetting.*	220
Fig A4.34 *'Mesh' of â for case (ii) using row forgetting.*	220
Fig A4.35 *'Mesh' of ĉ for case (ii) using row forgetting.*	221
Fig A4.36 *Vertical cross-section of â and ĉ for case (ii) using row forgetting.*	221
Fig A4.37 *Horizontal cross-section of â and ĉ for case (ii) using row forgetting.*	222
Fig A4.38 *'Mesh' of â for case (iii) using row forgetting.*	222
Fig A4.39 *'Mesh' of ĉ for case (iii) using row forgetting.*	223
Fig A4.40 *Horizontal cross-section of â and ĉ for case (iii) using row forgetting.*	223
Fig A4.41 *'Mesh' of â for case (iv) using row forgetting.*	224
Fig A4.42 *'Mesh' of ĉ for case (iv) using row forgetting.*	224
Fig A4.43 *Vertical cross-section of â and ĉ for case (iv) using row forgetting.*	225
Fig A4.44 *Horizontal cross-section of â and ĉ for case (iv) using row forgetting.*	225
Fig A4.45 *'Mesh' of â for case (v) using general two-dimensional forgetting.*	226
Fig A4.46 *'Mesh' of ĉ for case (v) using general two-dimensional forgetting.*	226
Fig A4.47 *'Mesh' of â for case (v) using column forgetting.*	227
Fig A4.48 *'Mesh' of ĉ for case (v) using column forgetting.*	227
Fig A4.49 *'Mesh' of â for case (v) using the hybrid algorithm.*	222
Fig A4.50 *'Mesh' of ĉ for case (v) using the hybrid algorithm.*	228

Selective glossary of abbreviations and symbols

Models

MD Machine direction.

CD Cross-machine direction.

NSHP Non-symmetric half-plane.

SHP Symmetric half-plane.

SWP Symmetric wedged-plane.

ARMA Auto-regressive moving average.

CARMA Auto-regressive moving average with control. The two-dimensional CARMA model in the semi-infinite plane is given by

$$A(w^{-1}, z^{-1})y(m,n) = z^{-\nu}B(w^{-1}, z^{-1})u(m,n) + C(w^{-1}, z^{-1})e(m,n)$$

When the plane has finite width it is given by

$$A_m(w^{-1}, z^{-1})y(m,n) = z^{-\nu}B_m(w^{-1}, z^{-1})u(m,n) + C_m(w^{-1}, z^{-1})e(m,n)$$

where m runs between 1 and W. The multivariable CARMA model is given by

$$\underline{A}(z^{-1})\underline{y}(n) = z^{-\nu}\underline{\underline{B}}(z^{-1})\underline{u}(n) + \underline{\underline{C}}(z^{-1})\underline{e}(n)$$

(m, n) Co-ordinates representing position on the plane.

$y(m, n)$ Output at (m, n).

$u(m, n)$ Input at (m, n).

$e(m,n)$ Noise at (m,n) with variance σ^2.

w^{-1}, z^{-1} Horizontal and vertical backwards shift operators respectively.

ν Delay in the process.

$\left.\begin{array}{l} A(w^{-1}, z^{-1}) \\ B(w^{-1}, z^{-1}) \\ C(w^{-1}, z^{-1}) \end{array}\right\}$ Two-dimensional polynomials with parameters $\left\{\begin{array}{l} a_{i,j} \\ b_{i,j} \\ c_{i,j} \end{array}\right.$

$A_m(w^{-1}, z^{-1})$ Polynomial corresponding to $A(w^{-1}, z^{-1})$ but truncated to take into account finite edges.

$\underline{\underline{A}}(z^{-1})$ One-dimensional polynomial matrix.

$\prec$ Relation meaning 'lies in the NSHP past with respect to'.

$\stackrel{j}{\leftrightarrows}$ Relation denoting equivalence between two-dimensional polynomials and entries in one-dimensional polynomial matrices (in some sense—see §3.3 for its precise definition).

$d(m)$ Constant offset in column m.

Δ Incremental term given by $\Delta = 1 - z^{-1}$.

Prediction and control

$T(w^{-1}, z^{-1})$ Desired closed-loop denominator in pole-assignment control.

$\hat{y}(m+k_1, n+k_2 | m, n)$ Least squares predictor of future output.

$\hat{\underline{y}}(n+k_2 | m, n)$ Multivariable least squares predictor.

$\left.\begin{array}{l} F^{k_1,k_2}(w^{-1}, z^{-1}) \\ G^{k_1,k_2}(w^{-1}, z^{-1}) \end{array}\right\}$ Polynomial solutions of the two-dimensional partition equation.

$\left.\begin{array}{l} \Phi^{k_1,k_2}(w^{-1}, z^{-1}) \\ \Gamma^{k_1,k_2}(w^{-1}, z^{-1}) \end{array}\right\}$ Finite-order polynomials corresponding to $\left\{\begin{array}{l} F^{k_1,k_2}(w^{-1}, z^{-1}) \\ G^{k_1,k_2}(w^{-1}, z^{-1}) \end{array}\right.$

$F_{m,k_1,k_2}(w^{-1}, z^{-1})$ Polynomial of future noise in the finite edge case.

$\underline{\underline{F}}_{k_2}^m(z^{-1})$ Polynomial matrix of future noise in the multivariable case.

$\left. \begin{array}{l} P(w^{-1}, z^{-1}) \\ Q'(w^{-1}, z^{-1}) \\ R(w^{-1}, z^{-1}) \end{array} \right\}$ Weighting polynomials for generalised minimum variance control.

$\phi(m, n)$ Pseudo-output for generalised minimum variance control.

$r(m, n)$ Setpoint.

$\left. \begin{array}{l} X_{m,0,\nu}(w^{-1}, z^{-1}) \\ Y_{m,0,\nu}(w^{-1}, z^{-1}) \\ Z_{m,0,\nu}(w^{-1}, z^{-1}) \end{array} \right\}$ Controller polynomials in the finite edge case.

Parameter estimation

RLS Recursive least squares.

RELS Recursive extended least squares.

AML Approximate maximum likelihood.

θ True parameter vector.

$\hat{\theta}(m, n)$ Vector of parameter estimates.

$\phi(m, n)$ Data vector.

ϵ (A priori) prediction error.

η Residual (a posteriori prediction error).

$P(m, n)$ Covariance matrix.

$Q(m, n)$ Information matrix.

$R(m, n)$ Vector given by $R(m, n) = \sum \phi^T(m, n) y(m, n)$. N.B. $P(m, n)$, $Q(m, n)$ and $R(m, n)$ are generally subscripted to denote data collection over certain regions of the plane. See for example Equations 4.2.2 to 4.2.5.

λ, μ Forgetting factors.

$\mathcal{R}$ Transformation corresponding to pre-multiplication by an orthogonal matrix resulting in an upper triangular matrix.

$\mathcal{H}$ Householder transformation.

$\mathcal{G}$ Fast Givens transformation.

1 Introduction

1.1 Background

In this monograph we address the problems of control, estimation and self-tuning as they apply to two-dimensional dynamical systems. In particular, we consider discrete models of two-dimensional processes. Such a process y may be defined as a linear time-invariant function of two indices

$$y = y(m, n)$$

Interest in such processes stems from such diverse fields as image processing, modelling of partial differential equations, analysis of geophysical data or analysis of delay-differential systems. In the image-processing application for example, the indices m and n are pixel column and row coordinates respectively and $y(m, n)$ is the image intensity at that pixel. In general m will be a spatial coordinate while n may be either a spatial or temporal coordinate.

The examples of application areas given above are all concerned with signal processing. However a growing and technically very important range of two-dimensional problems exist which require feedback control. An example of this type of control problem will be cited at length in §1.2 (below). As a precursor to a consideration of such applications it will be useful to discuss previous directions in two-dimensional control theory. Two-dimensional control has previously been approached from a predominantly systems theoretical point of view. This has two main branches, neatly encapsulated in the companion papers [Morf et al. 1977] and [Kung et al. 1977], taking an input-output transfer function approach and a state-space approach respectively.

Two-dimensional rational transfer functions were first studied in [Ozaki and Kadami 1960] and [Ansell 1964] and the main thrust of two-dimensional input-output based research by control theorists continues within a rational transfer function framework. Statistical approaches (for example [Whittle 1954]) have likewise taken an input-output model in the form of two-dimensional ARMA (auto-regressive moving-average) processes.

When applied to control, the transfer function approach has concentrated on the pole-assignment problem [Levy 1981], [Sebek 1985].

In the state-space formulation a major initial preoccupation of researchers concerned establishing standard forms for two-dimensional state-space models. Among the most commonly considered two-dimensional state-space models are those discussed in [Roesser 1975] and [Fornasini and Marcesini 1978]. It can be shown that there is a strong relationship between the different state-space models [Kaczorek 1985] and also between state-space and transfer function models [Kung et al. 1977]. More headway has been made in control using the state-space approach than using the transfer function approach. Results include stabilisation by static feedback [Fadali and Gnanadekaran 1989], [Lu and Lee 1985], dynamic feedback [Lee and Lu 1985], [Paraskevopoulos and Kosimdou 1981], minimum energy control [Kaczorek and Klamka 1986] and more recently optimal control [Lee and You 1988], [Li and Fadali 1991], [Bisiacco and Fornasini 1991].

However the difficult algebra of two-dimensional systems has made the controller design and analysis problematic for practical applications. In most cases the optimal control problem cannot be solved using a finite number of local states [Bisiacco and Fornasini 1991]. Where processes that can be modelled as two-dimensional are controlled it is rare to find the theory used or even referred to (a notable exception is the modelling of multipass processes as two-dimensional [Edwards and Owens 1982], [Rogers and Owens 1989]).

As noted above the lack of applications of two-dimensional control is due primarily to the complexity of the underlying theory (multipass systems can be treated analytically since the corresponding two-dimensional representation in this case is particularly simple). The problems in applying two-dimensional theory to real processes are further exacerbated by the significant difficulties in obtaining models of two-dimensional dynamical systems in a form suitable for control. It is appropriate once again to mention multipass processes where the simplicity of the two-dimensional structure assumed allows models to be obtained easily.

Within this discussion the two requirements for the creation of an applicable body of control theory emerge. Namely:

(1) The ability to model and parametrise the underlying dynamical process in an appropriate manner.

1.1 Background

(2) The ability to design *simple* control strategies for the underlying process *based upon the model*.

One of the motivations for this work is to build upon previous work at the Control Systems Centre and develop a set of compatible algorithms and associated theory for:

(1) Modelling two-dimensional CARMA (controlled auto-regressive moving average) processes via parameter estimation techniques.

(2) Controller synthesis for two-dimensional CARMA models.

(3) Combined estimation and controller synthesis in the form of two-dimensional self-tuning control.

The linkage of estimation and control in a self-tuning form is we believe especially important for two-dimensional systems since the large number of parameters within a two-dimensional CARMA model make traditional off-line procedures unreliable and impractical. By the same token commercial trends are such that almost all industrial controllers contain a self- or auto-tuning capability.

Self-tuning control,in one dimension can be traced as far back as [Kalman 1958] but the modern interest was sparked by [Peterka 1970] and the minimum variance controller [Astrom and Wittenmark 1973]. Self-tuning controllers consist of a simple control law dependent on parameters which may be estimated on-line. Self-tuning controllers using more sophisticated control laws have been developed including generalised minimum variance control [Clarke and Gawthrop 1975], [Gawthrop 1977], pole-assignment control [Edmunds 1976], [Wellstead et al. 1979], [Wellstead and Sanoff 1981] and more recently predictive self-tuning controllers,eg [Peterka 1984], [Clarke et al. 1987], [Lelic and Zarrop 1987]. Multivariable minimum variance [Borisson 1979] and generalised minimum variance [Koivo 1980] self-tuning controllers have also been developed.

The first self-tuning algorithms for two-dimensional processes were those developed for prediction, smoothing and deblurring [Caldas-Pinto 1983], [Wagner and Wellstead 1990], [Wagner 1987]. The philosophy here was to exploit the duality between prediction and control (see for example [Whittle 1983]) and apply one-dimensional control ideas to two-dimensional image-processing problems. In this monograph we attempt to bring the ideas full circle and apply them to the two-dimensional control problem.

1.2 Applications

The algorithms developed in this work are motivated by a particular class of two-dimensional process. In general, our primary motivating application is the generic task of manufacturing continuous webs (see Fig 1.2.1). Here a wide strip of material is to be produced length-ways and shows variations in its properties along both its length and breadth. Control is required in the two dimensions of length and breadth in order to regulate the properties of the finished web. The web-manufacturing process is generic in the sense that examples exist across a wide range of industries, including papermaking, plastic-film extrusion, cloth weaving and steel rolling. In the same spirit many continuous coating and plating processes have a two-dimensional nature and can be included within our definition of a web if we acknowledge that many manufactured webs are multi-layer.

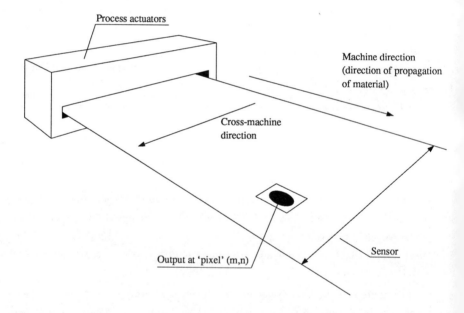

Fig 1.2.1 *Representation of the generic web-manufacturing process.*

Despite the clearly two-dimensional property of the general web-manufacturing pro-

1.2 Applications

cess (ie variations across both the length and breadth of the web), it will suit our purposes to refer to a specific web-forming process from time to time. Our preferred application is that of basis weight control in the papermaking industry. Here we give a brief description of the process and control problem with reference to Fig 1.2.2. For a more detailed account see [Smook 1982].

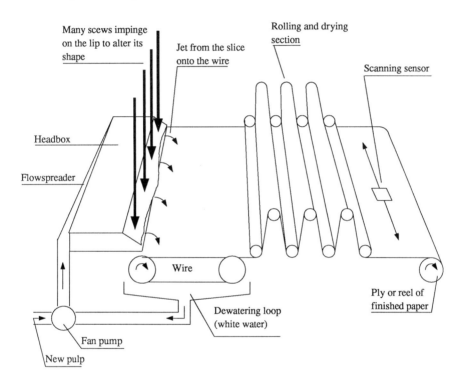

Fig 1.2.2 *Representation of the papermaking process.*

At the fan pump new pulp (a mix of water, wood fibre and other additives) is mixed with the white water and fed to the headbox via the flowspreader. At the headbox the pulp consists of about 99% water. It is fed at high pressure through the slice onto the wire section, where the pulp is carried on a continuously moving (up to $1300 m/min$)

felt. About 30% of the water is removed on the wire by drainage and fed back to the fan pump. Thereafter the web of paper is strong enough to be lifted off the wire and is fed through a series of rollers and dryers. Basis weight (dry weight per unit area), moisture and caliper (thickness) are all then monitored by a scanning sensor before the web is fed onto a reel. The direction in which the web moves is called the MD (machine direction) and that perpendicular to it the CD (cross-machine direction). Machines can be over $100m$ long and $6m$ wide.

For control purposes it is currently assumed that the variations in basis weight can be separated into machine direction variations, cross-machine direction variations and so-called residual variations [Wilhelm and Fjeld 1983]. These are considered to be independent and approximately equal. Machine direction variations are usually controlled with the speed of the fan pump. Cross-machine direction variations are then controlled by altering the profile of the slice. This is achieved using a series of evenly spaced screws which impinge on the top lip of the slice and are typically driven either by thermal expansion or harmonic drives. There may be as many as 200 such screws across the width of the machine. The residual variation is considered uncontrollable.

In the modeling of the formation process it has long been appreciated that there are strong interactions between the machine direction and cross-machine direction variations [Westermeyer 1987], [Holik et al. 1987]. Recently attempts have been made to take into account such interactions in the control design [Kastanakis and Lizr 1991]. However, the philosophy remains merely to control the steady-state profile of the cross-machine direction variations. Our contention is that such interactions can be accommodated within a two-dimensional model. Moreover we maintain that combined cross and machine direction control is not currently applied because the underlying two-dimensional control theory is not available. The practical motivation of this work is to provide the basis for such a theory.

1.3 Structure of the monograph

We begin (§2.1) by defining the two-dimensional control problem in the case where there are no edges to the data. This case allows us to apply the 'classical' two-dimensional algebraic results which we briefly survey in §2.2. We find (§2.3) that pole assignment is a poor candidate for a self-tuning control strategy. On the other hand we show (§2.4) that the infinite-order polynomials which occur in the two-dimensional prediction algorithm [Wagner and Wellstead 1990], [Wagner 1987] can be expressed as finite order rational polynomials. This leads naturally to an asymptotically optimal predictor (§2.5) and to the asymptotically optimal minimum variance and generalised minimum variance controllers. Their recursive nature allows them to have a much neater and more compact structure than the state-space optimal controllers referred to in §1.1.

In §3 we turn our attention to processes where the data are bound by finite edges. In §3.1 we rederive an optimal predictor and minimum variance and generalised minimum variance controllers. It turns out that these have a close relationship with their counterparts of §2. Their properties are demonstrated in some simple simulations in §3.2. However they may also be related to their multivariable counterparts [Borisson 1979], [Koivo 1980]. This relationship is explored in §3.3. In §3.4 we experiment with some more general structures for our two-dimensional process.

Two-dimensional parameter estimation is considered independently of any control or prediction algorithm in §4. The weighted least squares algorithm originally used for self-tuning prediction [Caldas-Pinto 1983], [Wagner and Wellstead 1990], [Wagner 1987] is re-examined (§4.1 and §4.2). An alternative forgetting strategy is also developed in §4.2. Conventional implementation techniques for forgetting factors are not satisfactory for the new strategy in its most general form. Some alternatives are developed in §4.3 and §4.4. Some special cases may be implemented using the more conventional techniques and these are considered in §4.5. In §4.6 the various forgetting schemes are compared in simulations.

In §5 we consider combining the generalised minimum variance controller of §3 with the parameter estimation techniques of §4 to form a self-tuning controller. Meanwhile we consider the two related problems, setpoint tracking (§5.2) and offset handling (§5.3). In each case if we permit specific structures to the model the problems are easily solved while the solution to the more general case is less simple.

1.4 Contribution

No originality is claimed for the results of §2.2. All the results have appeared previously in the literature, with the exception of Theorem 2.2.1 (which suggests a method for calculating the variance of a two-dimensional process with non-symmetric half-plane support). But this result is implicit from the translation discussed in §2.2.1 which was originally considered in [O'Connor and Huang 1981], [Levy 1981]. Similarly the pole-assignment problem (§2.3) has been considered in greater depth previously [Levy 1981], [Sebek 1985]. On the other hand the expression of the least squares predictor in terms of finite-order polynomials is believed to be new, as is the expression of asymptotically optimal minimum variance and generalised minimum variance control laws in terms of finite-order polynomials.

The application of two-dimensional ideas to processes of finite width is similar in philosophy to that of multipass processes [Edwards and Owens 1982], [Rogers and Owens 1989]. However once again the asymptotically optimal minimum variance and generalised minimum variance control laws are new. The results of §3 pave the way for more broad applications of two-dimensional models while the parallels drawn with multivariable models are novel and suggest new analysis techniques for two-dimensional processes.

In §4 the combination of weighted least squares familiar in image-processing applications with the recursive parameter estimation schemes of [Caldas-Pinto 1983], [Wagner and Wellstead 1990] and [Wagner 1987] allows the development of a versatile forgetting strategy. The usefulness of the strategy in various environments has yet to be proven, but already the analysis of §4.5 allows us to take into account the simple non-stationary behaviour expected at the edges of papermaking machines.

In §5 we develop a new class of self-tuning algorithms for two-dimensional processes. It is hoped that these will (in some modified form) prove applicable to real web processes and to these ends we consider some problems (setpoint tracking and offset handling) geared towards the practical control problem.

2 Prediction and control over the semi-infinite plane

In this chapter we consider the problem of two-dimensional prediction and control for processes generated over the full non-symmetric half-plane (NSHP). Although the assumption of such a global support is somewhat artificial it allows the results of 'classical' two-dimensional systems theory to be applied. It also allows us to develop generic algorithms which can be modified for specific applications later. We begin therefore, having defined the control problem (§2.1), by quoting the relevant results from two-dimensional systems theory (§2.2). A brief review and consideration of pole-assignment control for two-dimensional processes (§2.3) shows it to be an unwieldy technique and hence unsuitable for self-tuning; we do not propose any practical pole-assignment controller in this monograph. Instead it is shown that optimal-type strategies are algorithmically far easier to implement for two-dimensional processes due to a neat solution of the particular two-dimensional Diophantine equation involved. Hence in §2.4 we derive the least squares optimal predictor for a two-dimensional CARMA model and in §2.5 the minimum variance and generalised minimum variance controller.

2.1 The two-dimensional control problem

The two-dimensional CARMA process is given by

$$A(w^{-1}, z^{-1})y(m,n) = z^{-\nu} B(w^{-1}, z^{-1})u(m,n) + C(w^{-1}, z^{-1})e(m,n) \qquad (2.1.1)$$

where $y(m,n)$ is the output, $u(m,n)$ is the input and $e(m,n)$ is white noise. The two-dimensional polynomials A, B and C are given by

$$A(w^{-1}, z^{-1}) = \sum_{i=0}^{M}\sum_{j=0}^{N} a_{i,j} w^{-i} z^{-j} + \sum_{i=1}^{M}\sum_{j=1}^{N} a_{-i,j} w^{i} z^{-j}$$

$$B(w^{-1}, z^{-1}) = \sum_{i=0}^{M}\sum_{j=0}^{N} b_{i,j} w^{-i} z^{-j} + \sum_{i=1}^{M}\sum_{j=1}^{N} b_{-i,j} w^{i} z^{-j}$$

and

$$C(w^{-1}, z^{-1}) = \sum_{i=0}^{M}\sum_{j=0}^{N} c_{i,j} w^{-i} z^{-j} + \sum_{i=1}^{M}\sum_{j=1}^{N} c_{-i,j} w^{i} z^{-j}$$

The co-ordinates (m,n) represent position in the plane with respect to some arbitrary origin towards the bottom left-hand corner of the plane. The backwards shift operators w^{-1} and z^{-1} operate in the horizontal and vertical directions respectively. Thus for example

$$w^{-1} y(m,n) = y(m-1, n)$$

and

$$z^{-1} y(m,n) = y(m, n-1)$$

Notice that A, B and C are all NSHP (non-symmetric half-plane) causal, so the global past may be considered to be the semi-infinite NSHP.

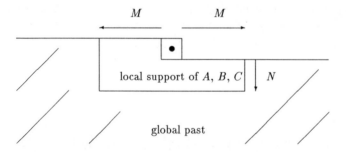

Fig 2.1.1 *Global and local supports for a two-dimensional NSHP process.*

The term $z^{-\nu}$ with ν some positive integer in Equation 2.1.1 corresponds to the delay in the process. We take it to be in the vertical direction since in general this will correspond to the temporal propagation of the process. For example the vertical direction of our model corresponds to the machine direction in papermaking, where the delay is between the scanning sensor at the dry end and the actuating screws at the wet end (see Fig 1.2.2). It is not hard to generalise our results to processes where the delay has a horizontal component as well; however we shall see in §3.4 (below) that for

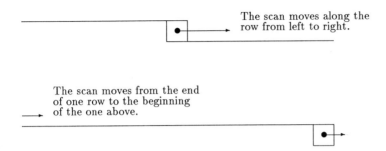

Fig 2.1.2 *The raster scan.*

practical applications in such a case it may be better to consider $B(w^{-1}, z^{-1})$ as spatially non-causal and retain the purely vertical delay.

We adopt the notation

$$(m_1, n_1) \prec (m_2, n_2)$$

if (m_1, n_1) lies in the NSHP past with respect to (m_2, n_2). That is to say

$$(m_1, n_1) \ \epsilon \ \{(m,n)| n < n_2 \text{ and } -\infty < m < \infty\} \cup$$
$$\{(m,n)| n = n_2 \text{ and } m < m_2\}$$

We may also say

$$y(m_1, n_1) \prec y(m_2, n_2)$$

with similar meaning. We define $\succ, \preceq$ and $\succeq$ similarly.

The data are read in a raster scan. The convention assumed here, corresponding to the NSHP configuration, is that the scan moves from left to right along each row, and upwards when going from the (theoretical) end of one row to the (theoretical) beginning of the next.

An admissible controller is one for which $u(m,n)$ is a function of $y(m-i, n-j)$ for $(m-i, n-j) \preceq (m,n)$ and of $u(m-i, n-j)$ for $(m-i, n-j) \prec (m,n)$.

2.2 Some results from two-dimensional systems theory

Here we present some results from the theory of two-dimensional systems and filters. Most such results are specifically for quarter-plane causal processes. We start therefore with a class of transformations which allow us to generalise such results to non-symmetric half-plane causal processes (§2.2.1). All the subsequent results which we quote are for quarter-plane causal processes, although in §2.2.4 we show that autocovariances are preserved under such a transformation. In §2.2.3 we quote the stability criterion for a two-dimensional transfer function and in §2.2.4 give various expressions for the autocovariance of an ARMA process. In §2.2.5 we show that a two-dimensional CARMA process may not always be stabilisable by a linear control law, even if $A(w^{-1}, z^{-1})$ and $z^{-\nu}B(w^{-1}, z^{-1})$ are factor coprime. This result stems from considerations of the zeros of two-dimensional polynomials discussed in §2.2.2.

2.2.1 Transforming from quarter-plane to non-symmetric half-plane causal processes

Suppose we have a NSHP causal operator given by

$$A(w^{-1}, z^{-1}) = \sum_{i=0}^{M}\sum_{j=0}^{N} a_{i,j} w^{-i} z^{-j} + \sum_{i=-M}^{-1}\sum_{j=1}^{N} a_{i,j} w^{-i} z^{-j}$$

We can transform this to

$$A'(w^{-1}, z^{-1}) = \sum_{i=0}^{M}\sum_{j=0}^{N} a_{i,j} w^{-i'} z^{-j'} + \sum_{i=-M}^{-1}\sum_{j=1}^{N} a_{i,j} w^{-i'} z^{-j'}$$

where

$$i' = k_1 i + k_2 j$$

and

$$j' = k_3 i + k_4 j$$

with k_1, k_2, k_3 and k_4 all integers. Such a transformation is discussed in both [O'Connor and Huang 1981] and [Levy 1981]. We require that

$$k_1 k_4 - k_2 k_3 \neq 0$$

though in [Levy 1981] this condition is further restricted to

$$k_1 k_4 - k_2 k_3 = \pm 1$$

It is shown in [O'Connor and Huang 1981] that under such a transformation stability is preserved. It is simple to show that properties such as variance are also preserved

2.2 Some results from two-dimensional systems theory

(§2.2.4). This becomes intuitively obvious if the transformation is interpreted as merely a change in the co-ordinates, or directions of the backwards shift operators.

In particular we may choose

$$k_1 = 1, \; k_2 = M, \; k_3 = 0 \text{ and } k_4 = 1$$

so that $A'(w^{-1}, z^{-1})$ is quarter-plane causal. Hence for analysis of properties such as stability and variance it is sufficient to consider quarter-plane causal processes.

2.2.2 Zeros and coprimeness

An important result for one-dimensional polynomials is the fundamental theorem of algebra. This may be interpreted as saying that any one-dimensional polynomial $a(z)$ can be expressed uniquely as a product of linear factors:

$$a(z) = \alpha_0 \prod_{i=1}^{n}(z - \alpha_i) \text{ with } \alpha_i \epsilon \mathcal{C}$$

where $\mathcal{C}$ denotes the complex numbers.

Such a result does not hold for two-dimensional polynomials. Although once again there is a unique factorisation to irreducible polynomials, these polynomials are no longer necessarily linear. This suggests immediately that the behaviour of zeros for two-dimensional polynomials is far richer than for one-dimensional polynomials.

We can define two kinds of coprimeness for two two-dimensional polynomials $A(w, z)$ and $B(w, z)$. We say A and B are factor coprime if there is no non-integer polynomial $P(w, z)$ that divides both A and B. We say A and B are zero coprime if there is no (w_0, z_0) such that

$$A(w_0, z_0) = B(w_0, z_0) = 0$$

Clearly

$$A, B \text{ zero coprime } \Rightarrow A, B \text{ factor coprime}$$

but unlike the one-dimensional case

$$A, B \text{ factor coprime } \not\Rightarrow A, B \text{ zero coprime}$$

We say

$$G(w, z) = \frac{B(w, z)}{A(w, z)}$$

is minimal if A and B are factor coprime. (w_0, z_0) is a nonessential singularity of G of the first kind if

$$A(w_0, z_0) = 0 \text{ and } B(w_0, z_0) \neq 0$$

while if G is minimal then (w_0, z_0) is a nonessential singularity of G of the second kind if

$$A(w_0, z_0) = B(w_0, z_0) = 0$$

2.2.3 Stability

A quarter-plane causal rational transfer function

$$\begin{aligned} G(w^{-1}, z^{-1}) &= \sum_{i=0}^{\infty} \sum_{j=0}^{\infty} g_{i,j} w^{-i} z^{-j} \\ &= \frac{B(w^{-1}, z^{-1})}{A(w^{-1}, z^{-1})} \\ &= \frac{\sum_{i=0}^{M} \sum_{j=0}^{N} b_{i,j} w^{-i} z^{-j}}{\sum_{i=0}^{M} \sum_{j=0}^{N} a_{i,j} w^{-i} z^{-j}} \end{aligned}$$

is said to be bounded-input bounded-output (bibo) stable if

$$\sum_{i=0}^{\infty} \sum_{j=0}^{\infty} |g_{i,j}| < \infty$$

Define the closed unit bi-sphere as

$$\overline{U}^2 = \{(w, z) : |w| \leq 1, |z| \leq 1, w, z \epsilon C\}$$

and the unit bi-disc

$$T^2 = \{(w, z) : |w| = 1, |z| = 1, w, z \epsilon C\}$$

Then we can say [Goodman 1977] that when G is minimal:

- If $A(w, z)$ is non-zero for all $(w, z) \epsilon \overline{U}^2$ then G is stable.
- If $A(w, z)$ is zero for some $(w_0, z_0) \epsilon \overline{U}^2 - T^2$ then G is unstable.

2.2 Some results from two-dimensional systems theory

- If $A(w, z)$ is zero for some $(w_0, z_0) \in T^2$ where the singularity at (w_0, z_0) is of the first kind then G is unstable.

- If $A(w, z)$ is non-zero for all $(w, z) \in \overline{U}^2 - T^2$ but $A(w, z)$ is zero for some $(w_0, z_0) \in T^2$ where all such zeros are of the second kind then G *may or may not* be stable. It can be shown [Goodman 1977] that the stability then depends on both A and B.

2.2.4 Autocovariance of an ARMA process

Consider the usual stable quarter-plane causal ARMA process given by

$$\begin{aligned}
y(m, n) &= \frac{C(w^{-1}, z^{-1})}{A(w^{-1}, z^{-1})} e(m, n) \\
&= H(w^{-1}, z^{-1}) e(m, n) \\
&= \sum_{i=0}^{\infty} \sum_{j=0}^{\infty} h_{i,j} e(m - i, n - j)
\end{aligned}$$

Its autocovariance is given [Hwang 1981] by

$$\begin{aligned}
r(p, q) &= E\left[y(m, n) y(m - p, n - q)\right] \\
&= \sigma^2 \sum_{i=0}^{\infty} \sum_{j=0}^{\infty} h_{i,j} h_{i-p, j-q} \\
&= \frac{\sigma^2}{(2\pi j)^2} \oint_{|z|=1} \oint_{|w|=1} \frac{C(w, z) C(w^{-1}, z^{-1})}{A(w, z) A(w^{-1}, z^{-1})} w^p z^q \frac{dw}{w} \frac{dz}{z}
\end{aligned}$$

where

$$\sigma^2 = E e(m, n)^2$$

For a general algorithm for the solution of this double complex integral see [Hwang 1981].

Here we show that such autocovariances are preserved under the transformation considered in §2.2.1.

Theorem 2.2.1 *Let $y(m, n)$ be a stable ARMA process given by*

$$A(w^{-1}, z^{-1}) y(m, n) = C(w^{-1}, z^{-1}) e(m, n)$$

where

$$A(w^{-1}, z^{-1}) = \sum_{i=0}^{M} \sum_{j=0}^{N} a_{i,j} w^{-i} z^{-j} + \sum_{i=1}^{M} \sum_{j=1}^{N} a_{-i,j} w^{i} z^{-j}$$

and

$$C(w^{-1}, z^{-1}) = \sum_{i=0}^{M} \sum_{j=0}^{N} c_{i,j} w^{-i} z^{-j} + \sum_{i=1}^{M} \sum_{j=1}^{N} c_{-i,j} w^{i} z^{-j}$$

Define also $y'(m, n)$ such that

$$A'(w^{-1}, z^{-1}) y'(m, n) = C'(w^{-1}, z^{-1}) e'(m, n)$$

where

$$A'(w^{-1}, z^{-1}) = \sum_{i} \sum_{j} a'_{i,j} w^{-i} z^{-j}$$

$$C'(w^{-1}, z^{-1}) = \sum_{i} \sum_{j} c'_{i,j} w^{-i} z^{-j}$$

with

$$a'_{k_1 i + k_2 j, k_3 i + k_4 j} = a_{i,j}$$

$$c'_{k_1 i + k_2 j, k_3 i + k_4 j} = c_{i,j}$$

and with k_1, k_2, k_3 and k_4 all integers satisfying

$$k_1 k_4 - k_2 k_3 = \pm 1$$

Define also

$$r(i, j) = E y(m, n) y(m - i, n - j)$$

$$r'(i, j) = E y'(m, n) y'(m - i, n - j)$$

Then

$$r'(k_1 i + k_2 j, k_3 + k_4 j) = r(i, j)$$

2.2 Some results from two-dimensional systems theory

Proof See Appendix 1.

Corollary 1 *Putting $i = j = 0$ gives*

$$\text{var } y' = \text{var } y$$

Corollary 2 *Putting $k_1 = 1$, $k_2 = M$, $k_3 = 0$ and $k_4 = 1$ ensures that y' is quarter-plane causal.*

Hence given an NSHP-causal ARMA process we can transform the co-ordinates to obtain an expression for the autocovariance function in terms of a quarter-plane causal transfer function.

2.2.5 The solution of two-dimensional Diophantine equations

Here we present two theorems from the algebra of two-dimensional polynomials. The discussion here is based on [Sebek 1983] and [Levy 1981]. For more traditional sources see [van der Waerden 1964] and [Fulton 1969].

Before we state the first we will state its one-dimensional equivalent. Specifically that for any two polynomials $A(z)$ and $B(z)$

$A(z)$ and $B(z)$ are coprime $\Leftrightarrow$ $\exists\ X(z)$ and $Y(z)$ such that

$$X(z)A(z) + Y(z)B(z) = 1$$

The significance of the two-dimensional result is that the coprimeness condition in one dimension translates to *zero* coprimeness in two dimensions. Hence:

Theorem 2.2.2 (Hilbert's Nullstellensatz) *If a polynomial $T(w,z)$ is zero at all common zeros of $A(w,z)$ and $B(w,z)$ then there exist an integer p and polynomials $X(w,z)$ and $Y(w,z)$ such that*

$$X(w,z)A(w,z) + Y(w,z)B(w,z) = T(w,z)^p$$

Corollary *Setting $T(w,z)$ to one gives the important result*

$A(w,z)$ and $B(w,z)$ are zero coprime

$\Leftrightarrow$ $\exists\ X(w,z)$ and $Y(w,z)$ such that

$$X(w,z)A(w,z) + Y(w,z)B(w,z) = 1$$

A value may be put on p; the result depends on Max Noether's Fundamental Theorem. Essentially it can be shown that the value of p depends on the *multiplicities* of the common zeros of $A(w,z)$ and $B(w,z)$. However the details are somewhat beyond the scope of this work (though the consequences are important). The interested reader is referred to [Sebek 1983], [Levy 1981] or [Fulton 1969] for details. The second theorem we state is Bezout's Theorem. But first we need to define the degree of a two-dimensional quarter-plane causal polynomial.

If

$$A(w,z) = \sum_{i=0}^{M}\sum_{j=0}^{N} a_{i,j} w^i z^j$$

then we define the degree of $A(w,z)$ as

$$\deg A(w,z) = MN$$

Then we may say:

Theorem 2.2.3 (Bezout's Theorem) *Given two polynomials $A(w,z)$ and $B(w,z)$, the number of points (w_0, z_0) such that*

$$A(w_0, z_0) = B(w_0, z_0) = 0$$

is

$$[\deg A(w,z)][\deg B(w,z)]$$

(though note that zeros may be repeated or at infinity).

In other words if we take any two polynomials $A(w,z)$ and $B(w,z)$ then almost surely they will not be zero coprime. Taking together Hilbert's Nullstellensatz and Bezout's Theorem (and also Max Noether's Fundamental Theorem) this may be interpreted as follows:

If we take any three polynomials $A(w,z)$, $B(w,z)$ and $T(w,z)$ then almost surely there will exist no polynomials $X(w,z)$ and $Y(w,z)$ such that we can solve

$$X(w,z)A(w,z) + Y(w,z)B(w,z) = T(w,z) \tag{2.2.1}$$

2.3 Pole-assignment control

However, if we *choose* $T(w, z)$ such that it shares the common zeros of $A(w, z)$ and $B(w, z)$ (and if common zeros are repeated then $T(w, z)$ must include these zeros 'a sufficient number of times') then we can solve an equation such as Equation 2.2.1.

Equation 2.2.1 is the two-dimensional Diophantine equation and we will need to solve different forms of it for pole-assignment control, least squares prediction and minimum variance type control strategies. The above result will thus be seen to be fundamental to this monograph.

2.3 Pole-assignment control

Suppose given our usual two-dimensional CARMA process

$$A(w^{-1}, z^{-1})y(m, n) = z^{-\nu}B(w^{-1}, z^{-1})u(m, n) + C(w^{-1}, z^{-1})e(m, n)$$

we apply *any* linear control law

$$P(w^{-1}, z^{-1})u(m, n) + Q(w^{-1}, z^{-1})y(m, n) = 0 \qquad (2.3.1)$$

then the closed-loop output is given by

$$y(m, n) = \frac{C(w^{-1}, z^{-1})P(w^{-1}, z^{-1})}{T(w^{-1}, z^{-1})}e(m, n)$$

where

$$T(w^{-1}, z^{-1}) = A(w^{-1}, z^{-1})P(w^{-1}, z^{-1}) + z^{-\nu}B(w^{-1}, z^{-1})Q(w^{-1}, z^{-1}) \qquad (2.3.2)$$

The discussion of §2.2.5 implies that $A(w^{-1}, z^{-1})$ and $z^{-\nu}B(w^{-1}, z^{-1})$ will almost surely share common zeros, and that $T(w^{-1}, z^{-1})$ must share them also. Hence for *any* linear control law there are certain poles which will always appear in the closed-loop denominator. This is an important difference between two-dimensional processes and one-dimensional and has many profound consequences. For example if $A(w^{-1}, z^{-1})$ and $z^{-\nu}B(w^{-1}, z^{-1})$ share such a common zero that lies in the instability region then the closed-loop output will always be unstable[1] [Sebek 1985]. Furthermore it makes pole-assignment control for two dimensions an unwieldy and unattractive design.

The one-dimensional pole-assignment controller (in a self-tuning context) is introduced in [Edmunds 1976], [Wellstead et al. 1979] and [Wellstead and Sanoff 1981]. Given the one-dimensional CARMA process given by

$$A(z^{-1})y(t) = z^{-k}B(z^{-1})u(t) + C(z^{-1})e(t)$$

---
[1] Here we exclude (as in one-dimensional control) cancellation of B via the controller, as this leads to an unstable input process in the case of nonminimum phase zeros.

the technique is essentially to choose a $T(z^{-1})$ and solve the one-dimensional Diophantine equation

$$X(z^{-1})A(z^{-1}) + z^{-k}Y(z^{-1})B(z^{-1}) = T(z^{-1})$$

for X and Y. Then applying the controller

$$X(z^{-1})u(t) + Y(z^{-1})y(t) = 0$$

will give a closed-loop process with denominator T. Given certain restrictions on the degrees of A,B,X,Y and T the one-dimensional Diophantine equation may always be solved, provided A and B have no common factors.

In two dimensions we cannot apply such a strategy since, as we have seen, if we are given $A(w^{-1}, z^{-1})$ and $z^{-\nu}B(w^{-1}, z^{-1})$ and we then choose *arbitrarily* $T(w^{-1}, z^{-1})$ then in general there will be no solution satisfying Equation 2.3.2. Instead a pole-assignment strategy in two dimensions might be implemented as follows:

Algorithm 2.3.1 (Two-dimensional pole-assignment control)

(1) *Find the common zeros of $A(w^{-1}, z^{-1})$ and $z^{-\nu}B(w^{-1}, z^{-1})$ (with their multiplicities).*

(2) *Choose $T(w^{-1}, z^{-1})$ such that*

- *$T(w^{-1}, z^{-1})$ shares the common zeros (with sufficient multiplicities).*
- *$T(w^{-1}, z^{-1})$ is stable.*
- *$T(w^{-1}, z^{-1})$ has suitable dynamics.*

(3) *Solve Equation 2.3.2.*

(4) *Implement the controller given by Equation 2.3.1.*

In [Sebek 1983] there is an algorithm to solve Equation 2.3.2 in the case where there is guaranteed to be a solution. If there is no solution it generates a polynomial in one variable, say $\phi_1(w^{-1})$, such that

$$A(w^{-1}, z^{-1})X(w^{-1}, z^{-1}) + z^{-\nu}B(w^{-1}, z^{-1})Y(w^{-1}, z^{-1}) = \phi_1(w^{-1})$$

Thus we could use the algorithm of [Sebek 1983] with $T(w^{-1}, z^{-1})$ set to 1 and then factorise $\phi_1(w^{-1})$ to find the common zeros of $A(w^{-1}, z^{-1})$ and $z^{-\nu}B(w^{-1}, z^{-1})$ in step

2.4 Prediction

(1). We could then choose a new $T(w^{-1}, z^{-1})$ and use the algorithm of [Sebek 1983] again to perform step (3).

However the details of implementing the algorithm are not worked out for the general case. In particular we have not considered how to count the multiplicities of common zeros, nor how to choose $T(w^{-1}, z^{-1})$ such that it shares these zeros with sufficient multiplicity. In general it seems unlikely that such an algorithm may be used in a self-tuning context, though as an off-line design strategy it may still be valuable. Furthermore the strategy has strayed from the philosophy of its one-dimensional counterpart. $T(w^{-1}, z^{-1})$ is no longer chosen arbitrarily but tailored to the characteristics of $A(w^{-1}, z^{-1})$ and $z^{-\nu} B(w^{-1}, z^{-1})$.

As an example consider the process given by

$$\left(1 - \tfrac{1}{2}w^{-1} - z^{-1}\right) y(m, n) = z^{-1} \left(1 + 2w^{-1}\right) u(m, n) + e(m, n)$$

Both $A(w^{-1}, z^{-1})$ and $B(w^{-1}, z^{-1})$ are inverse unstable so we must choose our controller with care. Step (1) of Algorithm 2.3.1 can be performed immediately by eye. The common zeros are at

$$(w^{-1}, z^{-1}) = (2, 0) \text{ and } (w^{-1}, z^{-1}) = (-\tfrac{1}{2}, \tfrac{5}{4})$$

Fortunately both these zeros lie in the stability region so we can choose an inverse stable $T(w^{-1}, z^{-1})$. For example we may choose

$$T(w^{-1}, z^{-1}) = (1 - \tfrac{1}{2}w^{-1})(1 - \tfrac{4}{5}z^{-1})$$

with solution

$$X(w^{-1}, z^{-1}) = 1$$

and

$$Y(w^{-1}, z^{-1}) = \tfrac{1}{5}$$

The cross-sections of these zeros in the real plane are shown in Fig 2.3.1.

2.4 Prediction

2.4.1 The two-dimensional prediction problem

Optimal prediction is a well-established technique of time-series analysis for one-dimensional processes (see for example [Astrom 1970] or [Wittenmark 1974]). A corresponding predictor for two-dimensional ARMA data has been developed in [Wagner

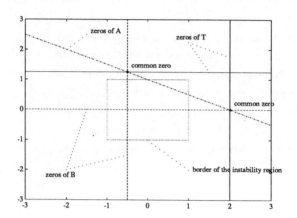

Fig 2.3.1 *Zeros of the polynomials $A(w^{-1},z^{-1})$, $z^{-1}B(w^{-1},z^{-1})$ and $T(w^{-1},z^{-1})$. Note that the loci of the zeros are in fact hyperplanes in four-space. Here we show their cross-sections in the real plane. Hence the instability region becomes the unit square (dotted line). The zeros of $A(w^{-1},z^{-1})$ lie on the line $1 - \frac{1}{2}w^{-1} - z^{-1} = 0$ (dashed-dotted line) while those of $z^{-1}B(w^{-1},z^{-1})$ lie on the two lines $z^{-1} = 0$ and $1 + 2w^{-1} = 0$ (dashed lines). Both sets of loci pass through the instability region, but their intersections at $(-\frac{1}{2},\frac{5}{4})$ and $(2,0)$ both lie outside. Thus we can choose an inverse stable $T(w^{-1},z^{-1})$ whose zeros lie on the two lines $1 - \frac{1}{2}w^{-1} = 0$ and $1 - \frac{4}{5}z^{-1} = 0$ (solid lines).*

and Wellstead 1990] and [Wagner 1987]. The predictor is given in recursive form with solution polynomials which turn out to be of infinite degree; a suboptimal solution is implemented by truncating these solution polynomials. Here we show that these infinite-order polynomials may be expressed as rational transfer functions of finite order. Thus by further polynomial manipulation a closed-form optimal two-dimensional predictor may be derived.

We adopt the philosophy that the predictor for a two-dimensional ARMA process is a special case of that for a CARMA process with $B(w^{-1},z^{-1})$ set to zero. Our model is then that of §2.1.

2.4 Prediction

Suppose

$$(m + k_1, n + k_2) \preceq (m, n + \nu)$$

Then the least squares predictor $\hat{y}(m + k_1, n + k_2 | m, n)$ may be defined as that predictor $y^*(m + k_1, n + k_2 | m, n)$ which minimises the cost function

$$E\left[y(m + k_1, n + k_2) - y^*(m + k_1, n + k_2 | m, n)\right]^2$$

given data up to and including (m, n).

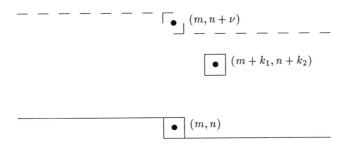

Fig 2.4.1 *Pixel to be predicted.*

As in one dimension [Wittenmark 1974], we construct the least squares predictor by truncating the expansion of C/A. Suppose

$$\frac{C(w^{-1}, z^{-1})}{A(w^{-1}, z^{-1})} = \sum_{j=0}^{\infty} \sum_{i=-jM}^{\infty} h_{i,j} w^{-i} z^{-j}$$

(This special structure results from the NSHP supports of $A(w^{-1}, z^{-1})$ and $C(w^{-1}, z^{-1})$.)
Then we define $F^{k_1, k_2}(w^{-1}, z^{-1})$ as

$$F^{k_1, k_2}(w^{-1}, z^{-1}) = \sum_{j=0}^{k_2-1} \sum_{i=-jM}^{\infty} h_{i,j} w^{-i} z^{-j} + \sum_{i=-k_2 M}^{k_1-1} h_{i, k_2} w^{-i} z^{-k_2} \tag{2.4.1}$$

with the convention that the first summation term is zero if $k_2 = 0$ and the second is zero if $-k_2 M > k_1 - 1$. Then by orthogonality for any predictor $y^*(m + k_1, n + k_2 | m, n)$ we have

$$E\left[y(m + k_1, n + k_2) - y^*(m + k_1, n + k_2 | m, n)\right]^2$$

$$\geq E\left[F^{k_1,k_2}(w^{-1},z^{-1})e(m+k_1,n+k_2)\right]^2$$

$$= \sigma^2 \sum_{j=0}^{k_2-1} \sum_{i=-jM}^{\infty} h_{i,j}^2 + \sigma^2 \sum_{i=-k_2M}^{k_1-1} h_{i,k_2}^2$$

Early work [Wellstead and Caldas-Pinto 1985], [Caldas-Pinto 1983] relied on the result that if the prediction is 'along the row', ie if $k_2 = 0$, then $F^{k_1,k_2}(w^{-1},z^{-1})$ has a finite number of terms. In other words the two-dimensional Diophantine equation

$$C(w^{-1},z^{-1}) = A(w^{-1},z^{-1})F^{k_1,0}(w^{-1},z^{-1}) + w^{-k_1}G^{k_1,0}(w^{-1},z^{-1}) \qquad (2.4.2)$$

has a solution with $F^{k_1,0}$ given by

$$F^{k_1,0}(w^{-1},z^{-1}) = \sum_{i=0}^{k_1-1} h_{i,0} w^{-i}$$

and with $G^{k_1,0}(w^{-1},z^{-1})$ causal. $G^{k_1,0}(w^{-1},z^{-1})$ may be determined from the relationship

$$G^{k_1,0}(w^{-1},z^{-1}) = w^{k_1}\left[C(w^{-1},z^{-1}) - A(w^{-1},z^{-1})F^{k_1,0}(w^{-1},z^{-1})\right]$$

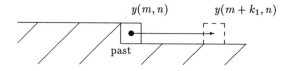

Fig 2.4.2 *Prediction 'along the row'.*

But if we wish to predict 'away from the row' we encounter problems. The corresponding Diophantine equation is still straightforward to construct, viz:

$$C(w^{-1},z^{-1}) = A(w^{-1},z^{-1})F^{k_1,k_2}(w^{-1},z^{-1}) + w^{-k_1}z^{-k_2}G^{k_1,k_2}(w^{-1},z^{-1}) \qquad (2.4.3)$$

However the solutions are no longer finite, since the partition gives $F^{k_1,k_2}(w^{-1},z^{-1})$ with support taking the form given in Equation 2.4.1.

In [Wagner and Wellstead 1990] and [Wagner 1987] algorithms are developed for prediction away from the line (ie with $k_2 > 0$) whereby truncated forms of $F^{k_1,k_2}(w^{-1},z^{-1})$ and $G^{k_1,k_2}(w^{-1},z^{-1})$ are calculated and then used directly. Results using these algorithms are good, but the necessity of truncation means that the prediction is non-optimal.

2.4.2 A closed form for the least squares predictor

Here we derive an optimal predictor based on the observation that $F^{k_1,k_2}(w^{-1}, z^{-1})$ and $G^{k_1,k_2}(w^{-1}, z^{-1})$ may both be expressed as finite-order rational transfer functions. To be specific:

Theorem 2.4.1 *There exist finite-order causal polynomials $\Phi^{k_1,k_2}(w^{-1}, z^{-1})$ and $\Gamma^{k_1,k_2}(w^{-1}, z^{-1})$ such that*

$$F^{k_1,k_2}(w^{-1}, z^{-1}) = \frac{\Phi^{k_1,k_2}(w^{-1}, z^{-1})}{A(w^{-1}, 0)^{k_2}} \qquad (2.4.4)$$

and

$$G^{k_1,k_2}(w^{-1}, z^{-1}) = \frac{\Gamma^{k_1,k_2}(w^{-1}, z^{-1})}{A(w^{-1}, 0)^{k_2}} \qquad (2.4.5)$$

Proof *See Appendix 1.*

We present an algorithm (Algorithm 2.4.1) for calculating $\Phi^{k_1,k_2}(w^{-1}, z^{-1})$ in Appendix 2. Given $\Phi^{k_1,k_2}(w^{-1}, z^{-1})$ the polynomial $\Gamma^{k_1,k_2}(w^{-1}, z^{-1})$ may be obtained from the relationship

$$\Gamma^{k_1,k_2}(w^{-1}, z^{-1}) = $$
$$w^{k_1} z^{k_2} \left[C(w^{-1}, z^{-1}) \left[A(w^{-1}, 0) \right]^{k_2} - A(w^{-1}, z^{-1}) \Phi^{k_1,k_2}(w^{-1}, z^{-1}) \right]$$

Hence we may configure a predictor which both is optimal and has a closed form.

Theorem 2.4.2 *The least squares predictor for the two-dimensional CARMA process of §2.1 is given recursively by*

$$C(w^{-1}, z^{-1}) \left[A(w^{-1}, 0) \right]^{k_2} \hat{y}(m + k_1, n + k_2 | m, n) = $$
$$B(w^{-1}, z^{-1}) \Phi^{k_1,k_2}(w^{-1}, z^{-1}) u(m + k_1, n + k_2 - \nu) + \Gamma^{k_1,k_2}(w^{-1}, z^{-1}) y(m, n)$$

$$(2.4.6)$$

Proof *See Appendix 1.*

The prediction error is then given by

$$\epsilon(m+k_1, n+k_2) = y(m+k_1, n+k_2) - \hat{y}(m+k_1, n+k_2|m,n)$$

$$= F^{k_1,k_2}(w^{-1}, z^{-1})e(m+k_1, n+k_2)$$

$$= \frac{\Phi^{k_1,k_2}(w^{-1}, z^{-1})}{[A(w^{-1},0)]^{k_2}} e(m+k_1, n+k_2)$$

Let V_{k_1,k_2} be the prediction-error variance. This may be calculated by transforming to a quarter-plane causal process and then performing the double integral as in [Hwang 1981] (see §2.2.4). Alternatively the special form for the prediction error allows us to calculate its variance in a more simple manner. We may say:

Theorem 2.4.3 *The variance of the prediction error V_{k_1,k_2} may be expressed as*

$$V_{k_1,k_2} = \sum_{\lambda=0}^{k_2-1} var\ y_\lambda(t) + \sum_{i=-\nu M}^{k_1-1} h_{i,k_2}^2 \qquad (2.4.7)$$

where

$$var\ y_\lambda(t) = \frac{1}{2\pi j} \oint_{|w|=1} \left[\frac{1}{[A(w^{-1},0)A(w,0)]^{k_2}} \frac{1}{w} \sum_{i=-\lambda M}^{k_2 M} \phi_{i,\lambda} w^{-i} \sum_{i=-\lambda M}^{k_2 M} \phi_{i,\lambda} w^i \right] dw$$

(2.4.8)

Proof *See Appendix 1.*

In other words the variance of the prediction error can be expressed as the sum of a finite number of one-dimensional integrals plus a further finite number of terms.

2.4.3 An algebraic interpretation

These results may be interpreted in terms of the results of §2.2. For prediction along the line in the special case where $k_2 = 0$ we wish to solve the two-dimensional Diophantine Equation 2.4.2. Note that this equation involves NSHP-causal polynomials. We can use a transform of the type introduced in §2.2.1 to convert this to an equation in quarter-plane causal polynomials—specifically we translate

$$z^{-1} \to w^{-M-1}\zeta^{-1}$$

2.4 Prediction

so that

$$A(w^{-1}, z^{-1}) = \sum_{i=0}^{M}\sum_{j=0}^{N} a_{i,j} w^{-i} z^{-j} + \sum_{i=1}^{M}\sum_{j=1}^{N} a_{-i,j} w^{i} z^{-j}$$

$$\to \quad A'(w^{-1}, \zeta^{-1}) = \sum_{i=0}^{M}\sum_{j=0}^{N} a_{i,j} w^{-i-j(M+1)} \zeta^{-j} + \sum_{i=1}^{M}\sum_{j=1}^{N} a_{-i,j} w^{i-j(M+1)} \zeta^{-j}$$

The Diophantine equation becomes

$$C'(w^{-1}, \zeta^{-1}) = A'(w^{-1}, \zeta^{-1}) F'^{k_1, 0}(w^{-1}, \zeta^{-1}) + w^{-k_1} G'^{k_1, 0}(w^{-1}, \zeta^{-1})$$

where all the polynomials are quarter-plane causal. But setting $w^{-1} = 0$ gives

$$A'(0, \zeta^{-1}) = a_{0,0}$$

$$\neq 0$$

Hence $A'(w^{-1}, \zeta^{-1})$ and w^{-k_1} share no common zero, so by Hilbert's Nullstellensatz (Theorem 2.2.2) we can find a solution for $F'^{k_1, 0}(w^{-1}, \zeta^{-1})$ and $G'^{k_1, 0}(w^{-1}, \zeta^{-1})$. We can then translate these solution polynomials back to the NSHP support.

But when we wish to predict away from the line (ie when $k_2 \geq 1$) we have to solve the Diophantine Equation 2.4.3. We find that there is in general no finite-order solution because $A(w^{-1}, z^{-1})$ and $w^{-k_1} z^{-k_2}$ share common zeros. By transforming from NSHP to quarter-plane causal as before we find that the w^{-k_1} term results in no common zero, so the common zeros are determined by

$$\left\{ w_0^{-1}, z_0^{-1} \,|\, A(w_0^{-1}, 0) = 0 \text{ and } z_0^{-1} = 0 \right\}$$

Instead we solve the equation

$$C(w^{-1}, z^{-1}) \left[A(w^{-1}, 0)\right]^{k_2} = A(w^{-1}, z^{-1}) \Phi^{k_1, k_2}(w^{-1}, z^{-1})$$
$$+ w^{-k_1} z^{-k_2} \Gamma^{k_1, k_2}(w^{-1}, z^{-1})$$

We find that $C(w^{-1}, z^{-1}) \left[A(w^{-1}, 0)\right]^{k_2}$ includes the common zeros with sufficient multiplicities to allow a solution.

Both the recursive form for the predictor (Equation 2.4.6) and the expression for the variance of the prediction error (Equations 2.4.7 and 2.4.8) imply that $A(w^{-1}, 0)$ must be inverse stable for the prediction error to be bounded. It is simple to show [O'Connor and Huang 1981] that

$$A(w^{-1},0) \text{ inverse unstable} \Rightarrow A(w^{-1},z^{-1}) \text{ inverse unstable}$$

Hence for an ARMA model this result says no more than that $A(w^{-1},z^{-1})$ must be inverse stable to allow prediction, corresponding with the familiar one-dimensional result. However we will find that the result may have dire consequences for control. Heuristically we may observe that if $k_2 > 0$ then any dynamics caused by the $A(w^{-1},0)$ component of the regression polynomial $A(w^{-1},z^{-1})$ on future data are independent of current known data.

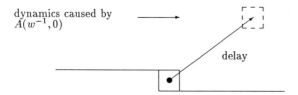

Fig 2.4.3 *Heuristic interpretation of the significance of $A(w^{-1},0)$.*

A further observation is that if

$$A(w^{-1},0) = 1$$

then the polynomials $F^{k_1,k_2}(w^{-1},z^{-1})$ and $G^{k_1,k_2}(w^{-1},z^{-1})$ will themselves have finite order.

2.5 Minimum variance and generalised minimum variance control

2.5.1 Minimum variance control

As in one dimension, the duality between least squares prediction and minimum variance control allows us to construct the minimum variance control law directly from the least squares predictor. Specifically, the minimum variance control law for the two-dimensional CARMA process of §2.1 is then given simply by letting

$$k_1 = 0 \text{ and } k_2 = \nu$$

2.5 Minimum variance and generalised minimum variance control

so that the partition of C/A becomes

$$C(w^{-1}, z^{-1}) = A(w^{-1}, z^{-1})F^{0,\nu}(w^{-1}, z^{-1}) + z^{-\nu}G^{0,\nu}(w^{-1}, z^{-1})$$

and then setting the least squares optimal predictor to zero. More formally, we may say:

Theorem 2.5.1 *The control law for the criterion*

$$\min_{u(m,n)} Ey(m, n+\nu)^2$$

is to set

$$B(w^{-1}, z^{-1})\Phi^{0,\nu}(w^{-1}, z^{-1})u(m, n) + \Gamma^{0,\nu}(w^{-1}, z^{-1})y(m, n) = 0$$

Proof *See Appendix 1.*

Thus we have expressed the minimum variance control law in terms of finite-order shift operators. In §3.2 (below) we show the results of its implementation in simulations.

The closed-loop output is

$$y(m, n) = \frac{\Phi^{0,\nu}(w^{-1}, z^{-1})}{[A(w^{-1}, 0)]^\nu} e(m, n)$$

while the input is

$$u(m, n) = \frac{-\Gamma^{0,\nu}(w^{-1}, z^{-1})}{[A(w^{-1}, 0)]^\nu B(w^{-1}, z^{-1})} e(m, n)$$

So just as in one dimension we require $B(w^{-1}, z^{-1})$ to be inverse stable. We have the further restriction that we require $A(w^{-1}, 0)$ to be inverse stable also. In fact the closed-loop output is unstable for *any* linear control law if $A(w^{-1}, 0)$ is inverse unstable. This follows from the discussion of §2.3 where we saw that the denominator of the closed-loop transfer function must include the common zeros of $A(w^{-1}, z^{-1})$ and $z^{-\nu}B(w^{-1}, z^{-1})$. Among these are the zeros of $A(w^{-1}, 0)$, so if this polynomial is inverse unstable then the closed-loop output will also be unstable (z^{-1} itself cannot be a factor of the denominator of the closed-loop transfer function if we are to have a causal control law).

2.5.2 Generalised minimum variance control

In the spirit of [Clarke and Gawthrop 1975] for the one-dimensional case we would like to generalise the minimum variance controller to some more suitable cost function. Specifically we would like to weight both the input and the output. Furthermore we would like to incorporate some setpoint $r(m,n)$. Let $P(w^{-1}, z^{-1})$, $Q'(w^{-1}, z^{-1})$ and $R(w^{-1}, z^{-1})$ be some general two-dimensional polynomials. Then we can say:

Theorem 2.5.2 *The control law for the criterion*

$$\min_{u(m,n)} E_{(m,n)} \left\{ [Py(m, n+\nu) - Rr(m,n)]^2 + [Q'u(m,n)]^2 \right\}$$

is equivalent to the minimum variance control law

$$\min_{u(m,n)} E\phi(m, n+\nu)^2$$

where $\phi(m,n)$ is the pseudo-output given by

$$\phi(m,n) = Py(m,n) + z^{-\nu}Qu(m,n) - z^{-\nu}Rr(m,n)$$

and Q is given by

$$Q(w^{-1}, z^{-1}) = \frac{Q'(0,0)}{B(0,0)} Q'(w^{-1}, z^{-1})$$

Proof See Appendix 1.

We may express $\phi(m,n)$ recursively as

$$A\phi = z^{-\nu}(PB + QA)u + PCe - z^{-\nu}Rr$$

Thus we may construct the generalised minimum variance controller in exactly the same way as we constructed the minimum variance controller. Specifically we partition

$$P(w^{-1}, z^{-1})C(w^{-1}, z^{-1}) = A(w^{-1}, z^{-1})F'(w^{-1}, z^{-1}) + z^{-\nu}G'(w^{-1}, z^{-1})$$

where

$$F'(w^{-1}, z^{-1}) = \sum_{i=0}^{\infty} f'_{i,0} w^{-i} + \sum_{j=1}^{\nu-1} \sum_{i=-\infty}^{\infty} f'_{i,j} w^{-i} z^{-j} + \sum_{i=-\infty}^{-1} f'_{i,\nu} w^{-i} z^{-\nu}$$

2.5 Minimum variance and generalised minimum variance control 31

and $G'(w^{-1}, z^{-1})$ is causal. Furthermore we may construct finite-order polynomials $\Phi'(w^{-1}, z^{-1})$ and $\Gamma'(w^{-1}, z^{-1})$ such that

$$\Phi'(w^{-1}, z^{-1}) = [A(w^{-1}, 0)]^\nu F'(w^{-1}, z^{-1})$$

and

$$\Gamma'(w^{-1}, z^{-1}) = [A(w^{-1}, 0)]^\nu G'(w^{-1}, z^{-1})$$

Then the control law is

$$\left\{B\Phi' + [A(w^{-1}, 0)]^\nu QC\right\} u(m, n) + \Gamma' y(m, n) = [A(w^{-1}, 0)]^\nu r(m, n)$$

In §3.2 (below) we show the results of the implementation of a modified form of this controller in simulation.

The closed-loop output is then

$$y(m, n) = \frac{z^{-\nu} BR}{AQ + BP} r(m, n) + \frac{B\Phi' + [A(w^{-1}, 0)]^\nu QC}{[A(w^{-1}, 0)]^\nu (AQ + BP)} e(m, n)$$

with corresponding input

$$u(m, n) = \frac{z^{-\nu} AR}{AQ + BP} r(m, n) - \frac{\Gamma'}{[A(w^{-1}, 0)]^\nu (AQ + BP)} e(m, n)$$

Thus for a stable closed-loop output we require both $A(w^{-1}, 0)$ and $AQ + BP$ to be inverse stable. The condition on $AQ + BP$ corresponds to the similar condition for the one-dimensional generalised minimum variance controller. However, unlike in one dimension we may not always be able to find polynomials P and Q such that $AQ + BP$ is inverse stable, even when A and B are factor coprime (see §2.3). The condition on $A(w^{-1}, 0)$ is specific to two-dimensional processes; $A(w^{-1}, 0)$ must be inverse stable for any linear controller to have stable closed-loop output (see §2.5.1).

3 Prediction and control over the plane of finite width

In this chapter we consider the problem of two-dimensional prediction and control for processes constrained by finite boundaries. In such cases the predictor derived in §2.4 and the controllers in §2.5 are suboptimal; the assumption of the semi-infinite global past corresponds to the assumption of a semi-infinite past in one-dimensional prediction and control. From a practical point of view if we can model the behaviour at the edges then the algorithms require only slight modifications to ensure optimality. In §3.1 we assume the local supports are simply truncated at the edges and develop the appropriate prediction and control algorithms. We also show that there is a very close link with the predictor of §2.4 and the controllers of §2.5. In §3.2 we illustrate these algorithms with some simulation results. In §3.3 we consider modelling our process as one-dimensional but multivariable (with the number of variables corresponding to the number of horizontal column positions). We show that there are close links between two-dimensional processes and multivariable processes. Finally in §3.3.7 we consider some more general types of behaviour at the edges. In §3.4 we consider the use of spatially non-causal local supports. We show that we can model more general types of behaviour using such supports and develop a controller for the case where the $B(w^{-1}, z^{-1})$ polynomial is spatially non-causal.

3.1 Prediction and control for case with edges

3.1.1 The process

For any real or simulated two-dimensional process we are likely to encounter edges. Furthermore we will usually demand that our control algorithm performs well at these edges. In this section we will consider how to modify our prediction and control algorithms to take into account one specific edge structure. In §3.3.7 (below) we consider more general types of behaviour at the edges.

In §2 we assumed a CARMA model where the horizontal index m ran from $-\infty$ to $+\infty$. Here we will assume m runs from 1 to some finite maximum W. Outside these

3.1 Prediction and control for case with edges

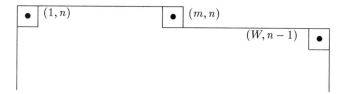

Fig 3.1.1 *Global support with edges.*

values we impose

$$y(m,n) = u(m,n) = e(m,n) = 0 \text{ for } m < 1 \text{ and } m > W \qquad (3.1.1)$$

An alternative way of picturing this is to assume that we truncate the local supports of A, B and C suitably at the edges. Thus our process becomes

$$A_m(w^{-1}, z^{-1})y(m,n) = z^{-\nu} B_m(w^{-1}, z^{-1})u(m,n) + C_m(w^{-1}, z^{-1})e(m,n) \qquad (3.1.2)$$

where

$$A_m(w^{-1}, z^{-1}) = \sum_{i=0}^{\min(M,m-1)} \sum_{j=0}^{N} a_{i,j} w^{-i} z^{-j} + \sum_{i=1}^{\min(M,W-m)} \sum_{j=1}^{N} a_{-i,j} w^{i} z^{-j} \qquad (3.1.3)$$

and similarly for B_m and C_m.

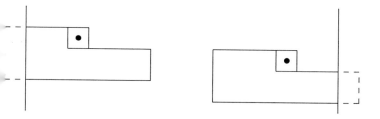

Fig 3.1.2 *The truncation of local supports.*

3.1.2 Prediction and minimum variance control

Suppose first we wish to construct the least squares optimal predictor

$$\hat{y}(m + k_1, n + k_2 | m, n) \text{ where } 1 \leq m + k_1 \leq W$$

We will obtain the predictor using the same principle that future noise is orthogonal to the known data; however we will construct the partition algorithmically rather than algebraically. Specifically put

$$A_m(w^{-1}, z^{-1}) = 1 + A'_m(w^{-1}, z^{-1})$$

and

$$C_m(w^{-1}, z^{-1}) = 1 + C'_m(w^{-1}, z^{-1})$$

so that

$$A'_m(w^{-1}, z^{-1})y(m, n) \prec y(m, n)$$

and

$$C'_m(w^{-1}, z^{-1})e(m, n) \prec e(m, n)$$

Then we may say

$$\begin{aligned}
y(m + k_1, n + k_2) &= -A'_{m+k_1}(w^{-1}, z^{-1})y(m + k_1, n + k_2) \\
&\quad + B_{m+k_1}(w^{-1}, z^{-1})u(m + k_1, n + k_2 - \nu) \\
&\quad + e(m + k_1, n + k_2) + C'_{m+k_1}(w^{-1}, z^{-1})e(m + k_1, n + k_2)
\end{aligned} \quad (3.1.4)$$

We may then exploit the fact that there is only a finite number of pixels (m_i, n_j) satisfying

$$(m, n) \prec (m_i, n_j) \prec (m + k_1, n + k_2)$$

Then we may substitute Equation 3.1.4 repeatedly into itself for $y(m_i, n_i)$ (note that the horizontal index on the polynomials will change) a finite number of times to obtain

$$\begin{aligned}
y(m + k_1, n + k_2) &= X_{m,k_1,k_2}(w^{-1}, z^{-1})y(m, n) \\
&\quad + Y_{m,k_1,k_2}(w^{-1}, z^{-1})u(m + k_1, n + k_2 - \nu) \\
&\quad + Z_{m,k_1,k_2}(w^{-1}, z^{-1})e(m, n) \\
&\quad + F_{m,k_1,k_2}(w^{-1}, z^{-1})e(m + k_1, n + k_2)
\end{aligned} \quad (3.1.5)$$

3.1 Prediction and control for case with edges   35

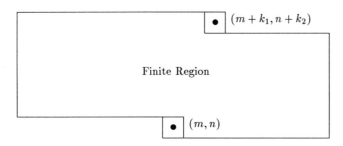

Fig 3.1.3 *The finite region between the predicted value and the known data.*

for some X_{m,k_1,k_2}, Y_{m,k_1,k_2}, Z_{m,k_1,k_2} and F_{m,k_1,k_2} where

$$F_{m,k_1,k_2}(w^{-1},z^{-1})e(m+k_1,n+k_2) \succ y(m,n)$$

and X_{m,k_1,k_2}, Y_{m,k_1,k_2} and Z_{m,k_1,k_2} are all causal.
Then by the usual orthogonality argument

$$E\left[y(m+k_1,n+k_2) - y^*(m+k_1,n+k_2|m,n)\right]^2 \geq$$
$$E\left[F_{m,k_1,k_2}(w^{-1},z^{-1})e(m+k_1,n+k_2)\right]^2$$

where $y^*(m+k_1,n+k_2|m,n)$ is any predictor given data up to (m,n) for the the future output $y(m+k_1,n+k_2)$.

For the special case

$$C_m(w^{-1},z^{-1}) = 1 \text{ for all } m$$

we have

$$Z_{m,k_1,k_2}(w^{-1},z^{-1}) = 0 \text{ for all } m$$

so we can implement directly the least squares optimal predictor

$$\hat{y}(m+k_1,n+k_2|m,n) = X_{m,k_1,k_2}(w^{-1},z^{-1})y(m,n)$$
$$+ Y_{m,k_1,k_2}(w^{-1},z^{-1})u(m+k_1,n+k_2-\nu)$$

For the more general case we can estimate the noise recursively using the process equation in the form

$$C_m(w^{-1},z^{-1})\hat{e}(m,n) = A_m(w^{-1},z^{-1})y(m,n) + B_m(w^{-1},z^{-1})u(m,n-\nu)$$

and then implement the predictor

$$\hat{y}(m+k_1, n+k_2|m,n) = X_{m,k_1,k_2}(w^{-1}, z^{-1})y(m,n)$$
$$+ Y_{m,k_1,k_2}(w^{-1}, z^{-1})u(m+k_1, n+k_2-\nu)$$
$$+ Z_{m,k_1,k_2}(w^{-1}, z^{-1})\hat{e}(m,n)$$

Since we assume $C(w^{-1}, z^{-1})$ to be inverse stable this estimate should converge to the true value (see §3.3.6 (below) for a brief discussion of this assumption) and our prediction error squared will converge to the optimal. Note that in a self-tuning context we may replace $\hat{e}(m,n)$ by the prediction error $\epsilon(m,n)$ or the residual $\eta(m,n)$ (see §4.1 (below) for their definitions in a two-dimensional context). Note also that in one-dimensional prediction and also our two-dimensional prediction without edges such an estimator for the noise is *implicit* in the algorithm; it is the presence of finite edges that forces us to estimate the noise *explicitly*.

Once again we may construct the minimum variance control law by letting

$$k_1 = 0 \text{ and } k_2 = \nu$$

and then setting the least squares predictor to zero. Specifically:

Theorem 3.1.1 *The control law for the criterion*

$$\min_{u(m,n)} Ey(m, n+\nu)^2$$

is to estimate the noise using

$$C_m(w^{-1}, z^{-1})\hat{e}(m,n) = A_m(w^{-1}, z^{-1})y(m,n) + B_m(w^{-1}, z^{-1})u(m, n-\nu)$$

(3.1.6)

and then to set

$$X_{m,0,\nu}(w^{-1}, z^{-1})y(m,n) + Y_{m,0,\nu}(w^{-1}, z^{-1})u(m,n) + Z_{m,0,\nu}(w^{-1}, z^{-1})\hat{e}(m,n)$$
$$= 0$$

Proof *See Appendix 1.*

Again we have a control law expressed in terms of finite-order shift operators. In §3.2 we show examples of its implementation and compare its performance with the generic minimum variance control law of §2.5.1.

3.1.3 The relationship with the case with no edges

We have now developed least squares optimal two-dimensional prediction and minimum variance control strategies for two different cases; specifically the case with no edges and the case with finite edges where we simply truncate the local supports. We go on to develop the generalised minimum variance control strategy for the second case in §3.1.4 but first we will pause to compare the two cases. Notice that in the first case our derivation was largely from an algebraic point of view whereas in the second case it was from an algorithmic point of view. Indeed two-dimensional algebraic theory implicitly assumes that we have no edges so it cannot strictly be applied to the second case *at all*. However we would like to use the two-dimensional algebraic theory for the second case (for example to analyse stability). It is with the aim of justifying such use that we compare the two cases here.

Suppose first of all that $A(w^{-1}, 0) = 1$.

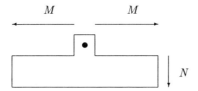

Fig 3.1.4 *Modified local support for $A(w^{-1}, z^{-1})$.*

Then for the case without edges it follows immediately from Equation 2.4.4 that the polynomial $F^{k_1, k_2}(w^{-1}, z^{-1})$ has finite order. In fact we may characterise $F^{k_1, k_2}(w^{-1}, z^{-1})$ as

$$F^{k_1,k_2}(w^{-1},z^{-1}) = \sum_{i=0}^{M} f_{i,0} w^{-i} + \sum_{j=1}^{k_2-1} \sum_{i=-jM}^{(j+1)M} f_{i,j} w^{-i} z^{-j}$$
$$+ \sum_{i=-k_2 M}^{\min((k_2+1)M, k_1-1)} f_{i,k_2} w^{-i} z^{-k_2}$$

Consider now the case with finite edges. We have already established in §3.1.2 that

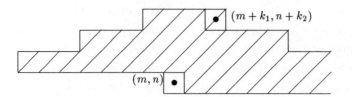

Fig 3.1.5 *Finite support for $F^{k_1,k_2}(w^{-1}, z^{-1})$ when $A(w^{-1}, 0) = 1$.*

$F_{m,k_1,k_2}(w^{-1}, z^{-1})$ is a finite-order polynomial. We have the further result that when

$$A(w^{-1}, 0) = 1$$

then provided

$$m + k_1 - k_2 M \geq 1$$

and

$$m + k_1 + k_2 M \leq W$$

then the derivation of $F_{m,k_1,k_2}(w^{-1}, z^{-1})$ is independent of the behaviour at the edges. We may then say:

Theorem 3.1.2 *Provided*

$$A(w^{-1}, 0) = 1$$

and

$$1 + k_2 M \leq m + k_1 \leq W - k_2 M \qquad (3.1.7)$$

we have the identity

$$F^{k_1,k_2}(w^{-1}, z^{-1}) = F_{m,k_1,k_2}(w^{-1}, z^{-1}) \qquad (3.1.8)$$

Proof *The result follows immediately by uniqueness.*

Since for each predictor the operator F determines the difference between the predicted value for y and its true value, it follows that given these restrictions on both

3.1 Prediction and control for case with edges

$A(w^{-1}, z^{-1})$ and (k_1, k_2) the two algorithms give the same prediction value. In other words sufficiently far from the edges the two algorithms are equivalent and interchangeable. A similar result follows for minimum variance control.

Although our restriction on $A(w^{-1}, 0)$ is quite severe it is useful since it allows us to reduce the computation in our prediction and control algorithms considerably. We will discuss the dynamical implications of the restriction in §3.4 (below). Note however that when

$$A(w^{-1}, 0) \neq 1$$

it follows that in general

$$F^{k_1, k_2}(w^{-1}, z^{-1}) \neq F_{m, k_1, k_2}(w^{-1}, z^{-1}) \text{ for all } m$$

In other words the predictor of §2.4 and minimum variance controller of §2.5 are suboptimal everywhere in this case. This suggests that although we developed a neat algebraic solution for the case when $A(w^{-1}, 0)$ is non-unity (Theorem 2.4.1) the result is never useful for the *implementation* of practical predictors and controllers (it remains useful of course as an analysis tool).

3.1.4 Generalised minimum variance control

As in §2.5 we would like to generalise the minimum variance control result to some more suitable cost function. Let $P_m(w^{-1}, z^{-1})$, $Q'_m(w^{-1}, z^{-1})$ and $R_m(w^{-1}, z^{-1})$ be the polynomials corresponding to the $P(w^{-1}, z^{-1})$, $Q'(w^{-1}, z^{-1})$ and $R(w^{-1}, z^{-1})$ introduced in §2.5 but truncated where necessary at the edges. Then:

Theorem 3.1.3 *The control law for the criterion*

$$\min_{u(m,n)} E_{(m,n)} \left\{ [P_m y(m, n + \nu) - R_m r(m, n)]^2 + [Q'_m u(m, n)]^2 \right\}$$

is equivalent to the minimum variance control law

$$\min_{u(m,n)} E\psi(m, n + \nu)^2$$

where $\psi(m, n)$ is the pseudo-output given by

$$\psi(m, n) = P_m y(m, n) + z^{-\nu} Q_m u(m, n) - z^{-\nu} R_m r(m, n)$$

and Q_m is given by

$$Q_m(w^{-1}, z^{-1}) = \frac{Q'_m(0, 0)}{B_m(0, 0)} Q'_m(w^{-1}, z^{-1})$$

Proof *See Appendix 1.*

Then we can construct the generalised minimum variance controller by repeatedly substituting from Equation 3.1.5 into the expression for ψ. In other words we construct the *finite edge* minimum variance controller for the *finite edge* pseudo-output ψ. The results of the implementation of such an algorithm are shown in §3.2 (below) and compared with those for the generic generalised minimum variance controller of §2.5.

Notice that if we are sufficiently far from the edges to prevent P_m, Q_m and R_m being truncated versions of P, Q and R respectively then

$$\psi(m,n) = \phi(m,n)$$

Hence we can set up the same kind of equivalence between the two versions of two-dimensional generalised minimum variance controllers as we set up for the predictors and minimum variance controllers in §3.1.3.

3.2 Simulations

Here we present some results of simulations of the minimum variance and generalised minimum variance control algorithms. Our aim is to demonstrate some simple examples of their implementation and performance when the process parameters are known. In §5 (below) we present results of their implementation in self-tuning form; we also examine in §5 some practical considerations which the controllers may encounter in applications.

In each simulation the process is given by

$$A_i(w^{-1}, z^{-1})y(m,n) = z^{-1}B_i(w^{-1}, z^{-1})u(m,n) + C(w^{-1}, z^{-1})e(m,n)$$

where

$$A_1(w^{-1}, z^{-1}) = 1 - 0.2w^{-1}z^{-1} - 0.4z^{-1} - 0.3wz^{-1}$$

$$A_2(w^{-1}, z^{-1}) = 1 - 0.6w^{-1}z^{-1} - 0.4z^{-1} - 0.5wz^{-1}$$

$$B_1(w^{-1}, z^{-1}) = 1 + 0.5w^{-1}z^{-1} + 0.1z^{-1} + 0.3wz^{-1}$$

$$B_2(w^{-1}, z^{-1}) = 1 + 0.5w^{-1}z^{-1} + 0.7z^{-1} + 0.4wz^{-1}$$

and

$$C(w^{-1}, z^{-1}) = 1 + 0.2w^{-1}z^{-1} - 0.3z^{-1} + 0.4wz^{-1}$$

3.2 Simulations

Note that A_1, B_1 and C are all inverse stable while A_2 and B_2 are inverse unstable. $e(m,n)$ is a white gaussian noise process with variance 1. $y(m,n)$ is generated over 50 by 50 pixels.

In the first simulation the process is given by

$$A_1(w^{-1}, z^{-1})y(m,n) = z^{-1}B_1(w^{-1}, z^{-1})u(m,n) + C(w^{-1}, z^{-1})e(m,n)$$

In Fig 3.2.1 we show the cumulative squared output for three cases: firstly the open-loop output, secondly for the generic minimum variance controller of §2.5.1 and thirdly for the minimum variance controller of §3.1.2 where the edges are considered. Note that the rows have been concatenated so that the 2500 pixels are presented as points on a straight line. If we had an ideal semi-infinite half-plane then minimum variance control should give a closed-loop output variance given by

$$Ey(m, n+1)^2 = E\left[\left(1 + 0.7wz^{-1}\right)e(m, n+1)\right]^2$$

$$= 1.49$$

This corresponds well with the actual mean squared output in the simulation. To the eye the two closed-loop cases are virtually indistinguishable; in fact the final value for the generic minimum variance controller is $3518 \approx 2500 \times 1.407$ while for when edges are considered the final value is $3506 \approx 2500 \times 1.403$.

In Fig 3.2.2 we show the difference between the two controllers and it is clear that they differ significantly only at the edges. We show both a two-dimensional 'mesh' of the difference and below the same data with the rows concatenated.

In the second simulation the process is open-loop unstable and given by

$$A_2(w^{-1}, z^{-1})y(m,n) = z^{-1}B_1(w^{-1}, z^{-1})u(m,n) + C(w^{-1}, z^{-1})e(m,n)$$

Again the two minimum variance controllers were applied and in Fig 3.2.3 we show the cumulative squared output for these two cases. This time if we had an ideal semi-infinite half-plane then minimum variance control should give a closed-loop output variance given by

$$Ey(m, n+1)^2 = E\left[\left(1 + 0.9wz^{-1}\right)e(m, n+1)\right]^2$$

$$= 1.81$$

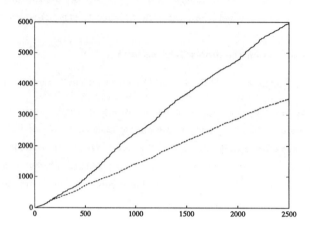

Fig 3.2.1 *Cumulative squared output for minimum variance control with $A_i = A_1$ and $B_i = B_1$. Solid line: open loop. Dashed line: generic minimum variance control. Dotted line (superimposed on dashed line): minimum variance control with edges considered.*

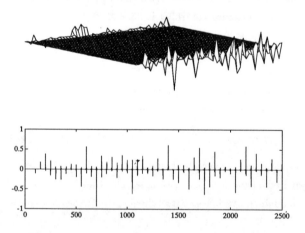

Fig 3.2.2 *The difference in outputs between the two minimum variance controllers with $A_i = A_1$ and $B_i = B_1$.*

Once again this corresponds well with the actual mean squared output in the simulation. This time the difference between the two graphs is just discernible. The generic minimum variance controller gives a final value of $4307 \approx 2500 \times 1.723$ which is reduced to $4230 \approx 2500 \times 1.692$ when edges are considered.

In Fig 3.2.4 as before we show the difference between the two controllers and again they differ significantly only at the edges.

In the third simulation the process is open-loop unstable with B also inverse unstable; it is given by

$$A_2(w^{-1}, z^{-1})y(m,n) = z^{-1}B_2(w^{-1}, z^{-1})u(m,n) + C(w^{-1}, z^{-1})e(m,n)$$

This time the two generalised minimum variance controllers were applied with

$$P(w^{-1}, z^{-1}) = Q(w^{-1}, z^{-1}) = 1$$

and with no setpoint: ie with

$$R(w^{-1}, z^{-1}) = 0$$

In other words this is the two-dimensional equivalent of the so-called 'lambda-controller' [Clarke and Gawthrop 1975]. In Fig 3.2.5 we show the cumulative squared output for the two cases. Again the two cases are virtually indistinguishable to the eye with final values $4961 \approx 2500 \times 1.985$ in both cases.

In Fig 3.2.6 as before we show the difference between the two controllers and once again they differ significantly only at the edges.

To conclude we have demonstrated that the minimum variance controller can both reduce the variance of a stable two-dimensional process and stabilise an unstable process when $B(w^{-1}, z^{-1})$ is inverse stable. We have also shown an example of the generalised minimum variance controller stabilising a two-dimensional process where $B(w^{-1}, z^{-1})$ is inverse unstable.

3.3 The Multivariable Connection

3.3.1 The process

In some respects our two-dimensional model for the case with finite edges described in §3.1.1 is unsatisfactory. We have already commented in §3.1.3 that two-dimensional algebraic results cannot strictly be applied to this case. Similarly we have the drawback that the coefficients of the operators such as $A_m(w^{-1}, z^{-1})$, $B_m(w^{-1}, z^{-1})$ and

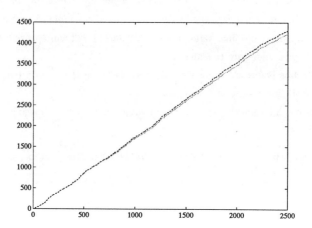

Fig 3.2.3 *Cumulative squared output for minimum variance control with $A_i = A_2$ and $B_i = B_1$. Dashed line: generic minimum variance control. Dotted line: minimum variance control with edges considered.*

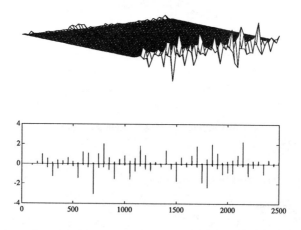

Fig 3.2.4 *The difference in outputs between the two minimum variance controllers with $A_i = A_2$ and $B_i = B_1$.*

3.3 The Multivariable Connection

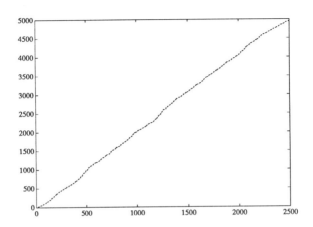

Fig 3.2.5 *Cumulative squared output for generalised minimum variance control with $A_i = A_2$ and $B_i = B_2$. Dashed line: generic generalised minimum variance control. Dotted line (superimposed on dashed line): generalised minimum variance control with edges considered.*

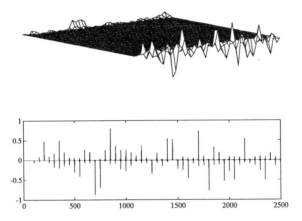

Fig 3.2.6 *The difference in outputs between the two generalised minimum variance controllers with $A_i = A_2$ and $B_i = B_2$.*

$C_m(w^{-1}, z^{-1})$ depend on the horizontal position m. Here we show how to model the process as multivariable so that the operators are fixed.

Consider for example the two-dimensional polynomial $A_m(w^{-1}, z^{-1})$.

$$A_m(w^{-1}, z^{-1}) = \sum_{i=0}^{\min(M,m-1)} \sum_{j=0}^{N} a_{i,j} w^{-i} z^{-j} + \sum_{i=1}^{\min(M,W-m)} \sum_{j=1}^{N} a_{-i,j} w^{i} z^{-j}$$

$$= \sum_{i=\max(-M,m-W)}^{\min(M,m-1)} a'_i(z^{-1}) w^{-i}$$

where

$$a'_i(z^{-1}) = \begin{cases} \sum_{j=1}^{N} a_{i,j} z^{-j} & \text{for } i < 0 \\ \sum_{j=0}^{N} a_{i,j} z^{-j} & \text{for } i \geq 0 \end{cases}$$

We can then say

$$A_m(w^{-1}, z^{-1}) y(m, n) = \sum_{i=\max(-M,m-W)}^{\min(M,m-1)} a'_i(z^{-1}) y(m-i, n)$$

If we define the W by 1 vector

$$\underline{y}(n) = \begin{bmatrix} y(1, n) \\ y(2, n) \\ \vdots \\ y(W, n) \end{bmatrix}$$

then we can see that $a'_i(z^{-1})$ operates on the $m - i$'th term in $\underline{y}(n)$. Hence if we construct the W by W matrix $\underline{A}$ whose (i, j)'th component is $a'_{j-i}(z^{-1})$ we can say

$$\begin{bmatrix} A_1(w^{-1}, z^{-1}) y(1, n) \\ \vdots \\ A_m(w^{-1}, z^{-1}) y(m, n) \\ \vdots \\ A_W(w^{-1}, z^{-1}) y(W, n) \end{bmatrix} = \begin{bmatrix} \sum_{i=\max(-M,1-W)}^{0} a'_i(z^{-1}) w^{-i} y(1, n) \\ \vdots \\ \sum_{i=\max(-M,m-W)}^{\min(M,m-1)} a'_i(z^{-1}) w^{-i} y(m, n) \\ \vdots \\ \sum_{i=0}^{\min(M,W-1)} a'_i(z^{-1}) w^{-i} y(m, n) \end{bmatrix}$$

3.3 The Multivariable Connection

$$= \begin{bmatrix} a'_0(z^{-1}) & a'_{-1}(z^{-1}) & \cdots & & 0 \\ a'_1(z^{-1}) & a'_0(z^{-1}) & \ddots & & \\ \vdots & \ddots & \ddots & \ddots & \vdots \\ & & \ddots & a'_0(z^{-1}) & a'_{-1}(z^{-1}) \\ 0 & & \cdots & a'_1(z^{-1}) & a'_0(z^{-1}) \end{bmatrix} \begin{bmatrix} y(1,n) \\ y(2,n) \\ \vdots \\ y(W,n) \end{bmatrix}$$

$$= \underline{A}(z^{-1})\underline{y}(n) \tag{3.3.1}$$

We may construct the polynomial matrices $\underline{B}(z^{-1})$ and $\underline{C}(z^{-1})$ and the vectors $\underline{u}(n)$ and $\underline{e}(n)$ similarly. Hence the W two-dimensional process equations

$$A_m(w^{-1}, z^{-1})y(m,n) = z^{-\nu} B_m(w^{-1}, z^{-1})u(m,n) + C_m(w^{-1}, z^{-1})e(m,n)$$

for m between 1 and W may be expressed as the single multivariable process equation

$$\underline{A}(z^{-1})\underline{y}(n) = z^{-\nu}\underline{B}(z^{-1})\underline{u}(n) + \underline{C}(z^{-1})\underline{e}(n) \tag{3.3.2}$$

This result establishes that we can model the dynamics of a two-dimensional process with edges as multivariable. This does not mean that there is a direct equivalence; for example the admissible control inputs in the two cases are different. In Appendix 3 we summarise the essential features of multivariable prediction, minimum variance control and generalised minimum variance control. Below we show how they may be modified in a simple manner for the two-dimensional case.

3.3.2 Prediction

The two-dimensional predictor of §3.1.2 $\hat{y}(m+k_1, n+k_2|m,n)$ is the best estimate of $y(m+k_1, n+k_2)$ given known data up to the point $y(m,n)$.

The multivariable predictor [Borisson 1979] given by $\underline{\hat{y}}(n+k_2|n)$ in §3.2 of Appendix 3 gives the best estimate of the whole row at $n+k_2$ given known data up to the end of the n'th row.

It is thus clear that the m'th entry of $\underline{\hat{y}}(n+k_2|n)$ is equal to $\hat{y}(m, n+k_2|W, n)$. Moreover, we know that

$$y(m, n+k_2) - \hat{y}(m, n+k_2|W, n) = F_{W, m-W, k_2}(w^{-1}, z^{-1})e(m, n+k_2)$$

and

$$\underline{y}(n+k_2) - \underline{\hat{y}}(n+k_2|n) = \underline{F}_{k_2}(z^{-1})\underline{e}(n+k_2)$$

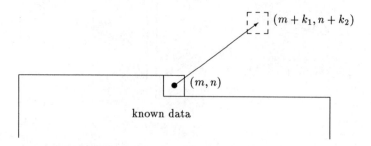

Fig 3.3.1 *Two-dimensional predictor.*

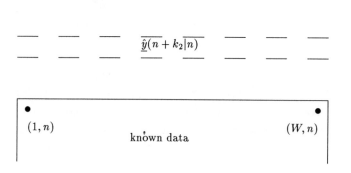

Fig 3.3.2 *Multivariable predictor.*

It follows by uniqueness that $F_{W,m-W,k_2}(w^{-1}, z^{-1})$ is equivalent to the m'th row of $\underline{\underline{F}}_{k_2}(z^{-1})$. That is to say, if the (i,j)'th term of $\underline{\underline{F}}_{k_2}(z^{-1})$ is $f_{i,j}(z^{-1})$ then

$$F_{W,m-W,k_2}(w^{-1}, z^{-1}) = \sum_{i=1}^{W} w^{i-m} f_{i,m}(z^{-1})$$

It is straightforward to modify the derivation of the multivariable predictor to take into account our NSHP knowledge of data. Specifically in §A3.2 of Appendix 3 we need only modify the partition of Equation A3.1 so that $\underline{\underline{F}}_k(z^{-1})\underline{e}(n+k)$ contains only future values in the NSHP sense. We shall label this adaptation of the predictor as $\hat{\underline{y}}(n+k|m,n)$. Note that the special case of the usual multivariable predictor according to this notation becomes $\hat{\underline{y}}(n+k|W,n)$.

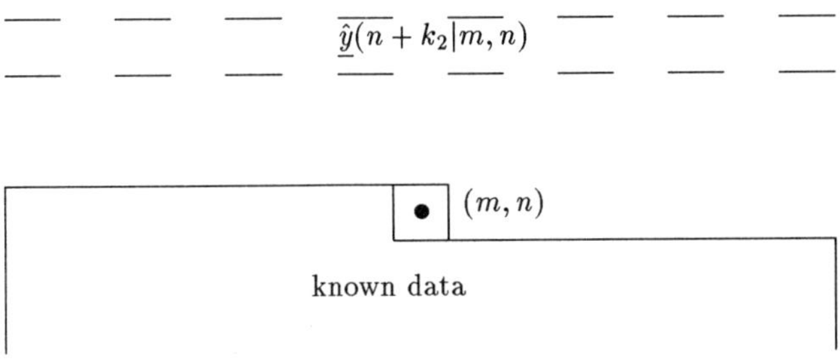

3.3 The Multivariable Connection

$$\overline{\hat{y}(n+k_2|m,n)}$$

$\bullet\ (m,n)$

known data

Fig 3.3.3 *The modified multivariable predictor.*

Hence in general $\hat{y}(m+k_1, n+k_2|m,n)$ is equal to the $(m+k_1)$'th entry of $\hat{\underline{y}}(n+k_2|m,n)$.

Let us now return to our original models. In the multivariable case if we express

$$\underline{y}(n) = z^{-\nu}\underline{\underline{A}}^{-1}\underline{\underline{B}}\,\underline{u}(n) + \underline{\underline{A}}^{-1}\underline{\underline{C}}\,\underline{e}(n)$$

then to form $\hat{\underline{y}}(n+k_2|m,n)$ we must partition $\underline{\underline{A}}^{-1}\underline{\underline{C}}$ into future and past parts. Let us adopt the notation

$$\underline{\underline{A}}^{-1}\underline{\underline{C}}\,\underline{e}(n+k_2) = \underline{\underline{F}}^m_{k_2}\underline{e}(n+k_2) + \underline{\underline{G}}^m_{k_2}\underline{e}(n) \qquad (3.3.3)$$

where the structures of $\underline{\underline{F}}^m_{k_2}$ and $\underline{\underline{G}}^m_{k_2}$ take into account our NSHP knowledge of data.

In the two-dimensional case we must find

$$y(m+k_1, n+k_2) - \hat{y}(m+k_1, n+k_2|m,n)$$
$$= F_{m,k_1,k_2}(w^{-1}, z^{-1})e(m+k_1, n+k_2)$$

The crucial observation is then that this polynomial $F_{m,k_1,k_2}(w^{-1}, z^{-1})$ is equivalent to the $(m+k_1)$'th row of the matrix $\underline{\underline{F}}^m_{k_2}(z^{-1})$. To be precise if the (i,j)'th term of $\underline{\underline{F}}^m_{k_2}(z^{-1})$ is $f^{m,k_2}_{i,j}$ then

$$F_{m,k_1,k_2}(w^{-1}, z^{-1}) = \sum_{i=1}^{W} w^{i-m-k_1} f^{m,k_2}_{i,m+k_1}(z^{-1}) \qquad (3.3.4)$$

3.3.3 Minimum variance control

In the two minimum variance control strategies we have different sets of admissible control strategies. For the two-dimensional controller with edges the input $u(m,n)$ is admissible if it is some function of

$$y(m-i, n-j) \text{ for } (m-i, n-j) \preceq (m,n)$$

and

$$u(m-i, n-j) \text{ for } (m-i, n-j) \prec (m,n)$$

Meanwhile for the multivariable controller [Borisson 1979] described in §A3.3 of Appendix 3 the input $\underline{u}(n)$ is admissible if it is some function of

$$\underline{y}(n-j) \text{ for } j \geq 0$$

and

$$\underline{u}(n-j) \text{ for } j > 0$$

To correspond with the two-dimensional case consider altering the set of admissible controllers for the multivariable case. Suppose we stipulate that an admissible control input $\underline{u}(n)$ is such that each element $u_m(n)$ is a function of $y_{m-i}(n)$ for $i \geq 0$, $\underline{y}(n-j)$ for $j \geq 1$, $u_{m-i}(n)$ for $i \geq 1$ and $\underline{u}(n-j)$ for $j \geq 1$. Then the only modification to the derivation of the control law of § A3.3 in Appendix 3 is that we perform the partition discussed in §3.3.2. Given the modified $\underline{\underline{F}}^m_\nu(z^{-1})$ the derivation follows exactly as in Appendix 3.

Furthermore from the error analysis of §3.3.2 we can conclude that given this modification the controllers become equivalent. But we are then faced with an apparent paradox in that the cost functions for the two controllers are not the same. Specifically for the two-dimensional controller we wish to minimise

$$\min_{u(m,n)} Ey(m, n+\nu)^2$$

(minimising a single output over a single input) whereas for the multivariable controller we wish to minimise

$$\min_{\underline{u}(n)} E\underline{y}^T(n+\nu)\underline{\underline{Q}}\,\underline{y}(n+\nu)$$

3.3 The Multivariable Connection

where $\underline{\underline{Q}}$ is any positive semi-definite matrix (so that the criterion is to minimise some quadratic function of a whole row of outputs over a whole row of inputs). Notice that choosing $\underline{\underline{Q}}$ to have its (m,m)'th entry equal to one and to be zero everywhere else does not render the criteria equal, since they are still minimising over different input sets. However we may say:

Theorem 3.3.1 *The controller which minimises*

$$\min_{u(m,n)} Ey(m, n+\nu)^2$$

also minimises

$$\min_{u(1,n),...,u(W,n)} E\sum_{i=1}^{W} y(i, n+\nu)^2$$

when applied successively (cf Appendix 2 in [Borisson 1979]).

Proof *See Appendix 1.*

In other words minimising $Ey(m,n+\nu)^2$ step by step at each control interval $u(m,n)$ gives the minimum value for the whole row $E\sum_{i=1}^{W} y(i, n+\nu)^2$ given any set of inputs $\{u(1,n),...,u(W,n)\}$.

It follows that the two-dimensional and multivariable cost functions are equivalent when

$$\underline{\underline{Q}} = \underline{\underline{I}}$$

and hence by uniqueness they are equivalent for all positive semidefinite $\underline{\underline{Q}}$. It may seem pedantic to have felt such a theorem necessary when the error analysis had already shown the two controllers to be equivalent. Its importance lies in that such a result does not necessarily hold for the generalised minimum variance controllers.

3.3.4 Generalised minimum variance control

A direct equivalence of the form established for minimum variance control does not exist between the generalised minimum variance control algorithm for two-dimensional data with edges derived in §3.1.4 and that for multivariable processes [Koivo 1980] described in §A3.4 of Appendix 3 even when modified as in §3.3.3. For the former the control criterion is

$$\min_{u(m,n)} E\left\{[P_m y(m, n+\nu) - R_m r(m,n)]^2 + [Q'_m u(m,n)]^2\right\}$$

whereas for the latter the control criterion is

$$\min_{\underline{u}(n)} E\left\{\left\|\underline{P}(z^{-1})\underline{y}(n+\nu) - \underline{R}(z^{-1})\underline{r}(n)\right\|^2 + \left\|\underline{Q}'(z^{-1})\underline{u}(n)\right\|^2\right\} \quad (3.3.5)$$

In [Koivo 1980] the $P(z^{-1})$ object is restricted to being a polynomial (as opposed to a polynomial matrix) only for algorithmic reasons. As we are not proposing to implement the controller in classical form we may consider a more general polynomial matrix $\underline{P}(z^{-1})$. Given the restrictions required for Equation 3.3.5 to describe the two-dimensional case, we may rewrite it as

$$\min_{u(1,n),\ldots,u(W,n)} E \sum_{i=1}^{W} \left\{[P_i(w^{-1}, z^{-1})y(i, n+\nu) - R_i(w^{-1}, z^{-1})r(i,n)]^2 + [Q'_i(w^{-1}, z^{-1})u(i,n)]^2\right\}$$

Thus the multivariable controller may be regarded as a form of predictive control for the two-dimensional case.

The control strategy for the former is to minimise (over a single input)

$$\psi = Py(m, n+\nu) + Qu(m,n) - Rr(m,n)$$

where

$$Q(w^{-1}, z^{-1}) = \frac{Q'(0,0)}{B(0,0)} Q'(w^{-1}, z^{-1})$$

whereas that for the latter it is to minimise (over a whole row of inputs)

$$\phi = \underline{P}\,\underline{y}(n+\nu) + \underline{Q}\,\underline{u}(n) - \underline{R}\,\underline{r}(n)$$

where

$$\underline{Q}(z^{-1}) = \left[\underline{B}^T(0)\right]^{-1} \underline{Q'}^T(0)\underline{Q'}(z^{-1})$$

Thus in general only if

$$B(w^{-1}, 0) = B(0,0)$$

and

$$Q'(w^{-1}, 0) = Q'(0,0)$$

3.3 The Multivariable Connection

so that

$$\underline{\underline{B}}(0) = \text{diag}\,(B(0,0),\ldots,B(0,0))$$

and similarly

$$\underline{\underline{Q'}}(0) = \text{diag}\,(Q'(0,0),\ldots,Q'(0,0))$$

do we have equivalence.

However we may still use multivariable theory to obtain a closed-loop description of the process. Applying Theorem 3.3.1 to our pseudo-output we may say that the two-dimensional generalised minimum variance controller minimises

$$\min_{u(1,n),\ldots,u(W,n)} E \sum_{i=1}^{W} \psi(i, n+\nu)^2$$

This then is equivalent to the multivariable minimum variance strategy for the pseudo-output

$$\underline{\psi}(n+\nu) = \underline{\underline{P}}\,\underline{y}(n+\nu) + \underline{\bar{\underline{Q}}}\,\underline{u}(n) - \underline{\underline{R}}\,\underline{r}(n) \tag{3.3.6}$$

where

$$\underline{\bar{\underline{Q}}}(z^{-1}) = \frac{Q'(0,0)}{B(0,0)}\underline{\underline{Q'}}(z^{-1})$$

We may combine Equation 3.3.6 with the process model (Equation 3.3.2) to obtain

$$\underline{\psi} = z^{-\nu}\left(\underline{\underline{P}}\,\underline{\underline{A}}^{-1}\underline{\underline{B}} + \underline{\bar{\underline{Q}}}\right)\underline{u} + \underline{\underline{P}}\,\underline{\underline{A}}^{-1}\underline{\underline{C}}\,\underline{e} - z^{-\nu}\underline{\underline{R}}\,\underline{r}$$

So if we partition

$$\underline{\underline{P}}\,\underline{\underline{A}}^{-1}\underline{\underline{C}} = \underline{\underline{F}} + z^{-\nu}\underline{\underline{G}}$$

for appropriate $\underline{\underline{F}}$ and $\underline{\underline{G}}$ then the control law is equivalent to

$$\left(\underline{\underline{P}}\,\underline{\underline{A}}^{-1}\underline{\underline{B}} + \underline{\bar{\underline{Q}}}\right)\underline{u} + \underline{\underline{G}}\,\underline{e} = \underline{\underline{R}}\,\underline{r}$$

giving closed-loop pseudo-output

$$\underline{\psi} = \underline{\underline{F}}\,\underline{e}$$

Again we may combine this with Equations 3.3.6 and 3.3.2 to obtain the closed-loop output

$$\left(\underline{\bar{\underline{Q}}}\,\underline{\underline{B}}^{-1}\underline{\underline{A}} + \underline{\underline{P}}\right)\underline{y} = z^{-\nu}\underline{\underline{R}}\,\underline{r} + \left(\underline{\bar{\underline{Q}}}\,\underline{\underline{B}}\,\underline{\underline{C}} + \underline{\underline{F}}\right)\underline{y} \tag{3.3.7}$$

We shall make use of this expression in §5 (below).

3.3.5 The link with the case without edges

So far we have explored the link between the two-dimensional case with finite edges and multivariable theory. We have shown that given certain modifications to the multivariable theory we obtain procedures whose results are asymptotically equivalent. Here we show that there is an algorithmic equivalence between the two-dimensional case *without* edges and the multivariable theory.

Recall from Theorem 2.4.2 that when

$$A(w^{-1}, 0) = 1$$

then the two-dimensional predictor without edges is given by

$$C(w^{-1}, z^{-1})\hat{y}(m + k_1, n + k_2 | m, n) =$$
$$B(w^{-1}, z^{-1})F^{k_1, k_2}(w^{-1}, z^{-1})u(m + k_1, n + k_2 - \nu) + G^{k_1, k_2}(w^{-1}, z^{-1})y(m, n)$$

where

$$C(w^{-1}, z^{-1}) = A(w^{-1}, z^{-1})F^{k_1, k_2}(w^{-1}, z^{-1}) + w^{-k_1} z^{-k_2} G(w^{-1}, z^{-1})$$

So we may say

$$\begin{aligned} C\hat{y}(m + k_1, n + k_2 | m, n) &= BF^{k_1, k_2} u(m + k_1, n + k_2 - \nu) \\ &+ \left(C - F^{k_1, k_2} A \right) y(m + k_1, n + k_2) \end{aligned} \quad (3.3.8)$$

Similarly, using Equation 3.3.3, we can express the modified multivariable predictor of §3.3.2 as the $m + k_1$'th entry of the vector

$$\hat{y}(n + k_2 | m, n) = \left[\underline{\tilde{C}}^m_{k_2}(z^{-1}) \right]^{-1} \left[\underline{\tilde{F}}^m_{k_2}(z^{-1}) \underline{B}(z^{-1}) \underline{u}(n + k_2 - \nu) + \underline{\tilde{G}}^m_{k_2}(z^{-1}) \underline{y}(n) \right] \quad (3.3.9)$$

where

$$\underline{C}(z^{-1}) = \underline{A}(z^{-1}) \underline{F}^m_{k_2} + z^{-k_2} \underline{G}^m_{k_2}$$

$\underline{\tilde{F}}^m_{k_2}(z^{-1})$ and $\underline{\tilde{G}}^m_{k_2}(z^{-1})$ are chosen (non-uniquely) such that

$$\underline{\tilde{F}}^m_{k_2}(z^{-1}) \underline{G}^m_{k_2}(z^{-1}) = \underline{\tilde{G}}^m_{k_2}(z^{-1}) \underline{F}^m_{k_2}(z^{-1})$$

with

$$\det \underline{\tilde{F}}^m_{k_2}(z^{-1}) = \det \underline{F}^m_{k_2}(z^{-1})$$

3.3 The Multivariable Connection

and

$$\underline{\tilde{F}}^m_{k_2}(0) = \underline{I}$$

Similarly we define

$$\underline{\tilde{C}}^m_{k_2}(z^{-1}) = \underline{\tilde{F}}^m_{k_2}(z^{-1})\underline{A}(z^{-1}) + z^{-k_2}\underline{\tilde{G}}^m_{k_2}(z^{-1})$$

so that

$$\underline{\tilde{C}}^m_{k_2}(z^{-1})\underline{F}^m_{k_2}(z^{-1}) = \underline{\tilde{F}}^m_{k_2}(z^{-1})\underline{C}(z^{-1})$$

It was noted above that the choice of $\underline{\tilde{F}}^m_{k_2}(z^{-1})$ is non-unique. One possibility would be to choose

$$\underline{\tilde{F}}^m_{k_2}(z^{-1}) = \underline{C}(z^{-1})\underline{F}^m_{k_2}(z^{-1})\left[\underline{C}(z^{-1})\right]^{-1}$$

so that

$$\underline{\tilde{C}}^m_{k_2}(z^{-1}) = \underline{C}(z^{-1})$$

and

$$\begin{aligned} z^{-k_2}\underline{\tilde{G}}^m_{k_2}(z^{-1}) &= \underline{\tilde{C}}^m_{k_2}(z^{-1}) - \underline{\tilde{F}}^m_{k_2}(z^{-1})\underline{A}(z^{-1}) \\ &= \underline{C}(z^{-1}) - \underline{C}(z^{-1})\underline{F}^m_{k_2}(z^{-1})\left[\underline{C}(z^{-1})\right]^{-1}\underline{A}(z^{-1}) \end{aligned}$$

Furthermore we may multiply both sides of Equation 3.3.9 by $\underline{C}(z^{-1})$ to obtain

$$\underline{C}\,\hat{\underline{y}}(n+k_2|m,n) = \underline{C}\,\underline{F}^m_{k_2}\left[\underline{C}\right]^{-1}\underline{B}\,\underline{u}(n+k_2-\nu) + \left[\underline{C} - \underline{C}\,\underline{F}^m_{k_2}\left[\underline{C}\right]^{-1}\underline{A}\right]\underline{y}(n+k_2) \quad (3.3.10)$$

Comparing Equations 3.3.8 and 3.3.10 it becomes clear that there is a structural correspondence between the two predictors. We can go further and say that given certain conditions there is a direct equivalence between the $m + k_1$'th row of the multivariable predictor above and the two-dimensional predictor.

Let us define the relation $\stackrel{j}{\sim}$ as follows. For the general two-dimensional polynomial

$$X(w^{-1}, z^{-1}) = \sum_{i=0}^{M}\sum_{j=0}^{N} x_{i,j} w^{-i} z^{-j} + \sum_{i=1}^{M}\sum_{j=1}^{N} x_{-i,j} w^{i} z^{-j}$$

and for the W by W polynomial matrix $\underline{\underline{X}}(z^{-1})$ with (i,j)'th element $\xi_{i,j}$ we will say

$$X(w^{-1}, z^{-1}) \stackrel{j}{\sim} \underline{\underline{X}}(z^{-1})$$

when

$$X(w^{-1}, z^{-1}) = \sum_{i=1}^{W} \xi_{j,i}(z^{-1}) w^{i-j}$$

Then

Theorem 3.3.2 *Suppose $A(w^{-1}, 0) = 1$. Then for W sufficiently large, and for both $m + k_1$ and $W - m - k_1$ sufficiently large*

$$C(w^{-1}, z^{-1}) \stackrel{m+k_1}{\sim} \underline{\underline{C}}(z^{-1})$$

$$F^{k_1, k_2}(w^{-1}, z^{-1}) B(w^{-1}, z^{-1}) \stackrel{m+k_1}{\sim} \underline{\underline{C}}(z^{-1}) \underline{\underline{F}}_{k_2}^{m}(z^{-1}) \left[\underline{\underline{C}}(z^{-1})\right]^{-1} \underline{\underline{B}}(z^{-1})$$

and

$$C(w^{-1}, z^{-1}) - F^{k_1, k_2}(w^{-1}, z^{-1}) A(w^{-1}, z^{-1}) \stackrel{m+k_1}{\sim}$$
$$\underline{\underline{C}}(z^{-1}) - \underline{\underline{C}}(z^{-1}) \underline{\underline{F}}_{k_2}^{m}(z^{-1}) \left[\underline{\underline{C}}(z^{-1})\right]^{-1} \underline{\underline{A}}(z^{-1})$$

Proof *See Appendix 1.*

Comparing Equations 3.3.8 and 3.3.10 we may interpret this theorem as saying that the two predictors perform the same operation away from the edges. A similar result is immediate for minimum variance control. Notice however that the implementation of the multivariable predictor is non-unique; the equivalence only holds for one implementation.

3.3.6 Reprise

We have shown that when there are edges the two-dimensional CARMA process may be modelled exactly as a multivariable process. Furthermore the least squares optimal predictor and the minimum variance control law correspond to their multivariable equivalents given certain modifications (although this does not extend to the generalised minimum variance controller).

We stress that we do not consider the multivariable forms suitable for implementation. This is partly because the number of parameters (W) will in general be large;

3.3 The Multivariable Connection

but also because, as we will see in §4 (below), we can exploit our two-dimensional formulation to estimate the parameters efficiently. Instead we can use the multivariable description to provide an exact analysis tool as opposed to the two-dimensional theory which can only be at best an approximation when there are edges. Note that the relationship between the two two-dimensional cases discussed in §3.1.3 and the relationship between the no-edge case and the multivariable theory discussed in §3.3.5 imply that the two-dimensional theory is often a good approximation.

To show where two-dimensional theory may be inadequate consider the following example. Suppose we have the two-dimensional process

$$A(w^{-1}, z^{-1})y(m,n) = e(m,n)$$

where

$$A(w^{-1}, z^{-1}) = \left(1 + \lambda w^{-1} z^{-1}\right)^2 \left(1 + \lambda w z^{-1}\right)$$

$A(w^{-1}, z^{-1})$ is inverse stable for $|\lambda| < 1$. Now consider its multivariable equivalent when there are edges. Put

$$a_0 = \lambda^2 z^{-2}$$

$$a_1 = 2\lambda z^{-1} + \lambda^3 z^{-3}$$

$$a_2 = 1 + 2\lambda^2 z^{-2}$$

$$a_3 = \lambda z^{-1}$$

Then if m runs between 1 and W we have

$$\underline{\underline{A}}_W(z^{-1}) = \begin{bmatrix} a_2 & a_3 & & & & & \\ a_1 & a_2 & a_3 & & & & \\ a_0 & a_1 & a_2 & a_3 & & & \\ & a_0 & a_1 & a_2 & a_3 & & \\ & & \ddots & \ddots & \ddots & \ddots & \\ & & & & a_0 & a_1 & a_2 & a_3 \\ & & & & & a_0 & a_1 & a_2 \end{bmatrix}$$

Let

$$\alpha_W(z^{-1}) = \det \underline{\underline{A}}_W(z^{-1})$$

Then we have

$$\alpha_1(z^{-1}) = a_2$$

$$\alpha_2(z^{-1}) = a_2^2 - a_1 a_3$$

$$\alpha_3(z^{-1}) = a_2\left(a_2^2 - a_1 a_3\right) - a_3\left(a_1 a_2 - a_0 a_3\right)$$

together with the recursive relationship

$$\alpha_W(z^{-1}) = a_2 \alpha_{W-1}(z^{-1}) - a_1 a_3 \alpha_{W-2}(z^{-1}) + a_0 a_3^2 \alpha_{W-3}(z^{-1}) \qquad (3.3.11)$$

Without analysing these determinants too deeply we may note that $\alpha_W(z^{-1})$ has the form

$$\alpha_W(z^{-1}) = \sum_{i=0}^{W} b_i \lambda^{2i} z^{-2i}$$

$$= \prod_{i=1}^{W} \left(1 + \beta_i z^{-2i}\right)$$

for some b_i and β_i. Then observing the highest order term in z^{-1} we know

$$\prod_{i=1}^{W} \beta_i = b_W \lambda^{2W}$$

so if $b_W \lambda^{2W} > 1$ then the determinant has at least one unstable root. From our recursive formula for α_W (Equation 3.3.11) we can see that

$$b_W = W + 1$$

so the process is unstable whenever

$$(W+1)\lambda^{2W} > 1$$

So for example if $\lambda = 0.9$ then the process is unstable at least for $W \leq 12$.

On the other hand the two-dimensional theory may often be a better guide to behaviour than the multivariable theory. Suppose now we have the process

$$A(w^{-1}, z^{-1}) y(m, n) = e(m, n)$$

where

$$A(w^{-1}, z^{-1}) = 1 + 2w^{-1} z^{-1}$$

3.3 The Multivariable Connection

Here $A(w^{-1}, z^{-1})$ is inverse unstable. Yet the equivalent multivariable matrix is given by

$$\underline{\underline{A}}(z^{-1}) = \begin{pmatrix} 1 & & & & \\ 2z^{-1} & 1 & & & \\ & 2z^{-1} & 1 & & \\ & & \ddots & \ddots & \\ & & & 2z^{-1} & 1 \end{pmatrix}$$

with

$$\det \underline{\underline{A}}(z^{-1}) = 1$$

Hence any implementation will in fact be stable. *However* the variance of any implementation, even for small W, will be very large and for any practical purposes the process could be considered unstable. This is an example of the two-dimensional theory giving a 'quick and dirty' approximate result which is not immediately apparent from the more formally correct multivariable theory.

Note that our two-dimensional algorithms assume the noise polynomial $C(w^{-1}, z^{-1})$ to be inverse stable. In one-dimensional single and multivariable theory this can be guaranteed. If our two-dimensional process is more correctly modelled as multivariable then it may occur that the multivariable matrix $\underline{\underline{C}}(z^{-1})$ is inverse stable while its two-dimensional equivalent $C(w^{-1}, z^{-1})$ is inverse unstable. However as the closed-loop response is determined by the multivariable matrices this should not cause practical problems.

3.3.7 Other models for behaviour at the edges

The edge conditions assumed in §3.1.1 constitute just one possibility among many. The condition expressed in Equation 3.1.1 that y, u and e are all zero off the edges of the plane seems reasonable for the control problem (though note that in image processing we may often have the problem that y is generated beyond the region of data with which we are concerned). However the assumption that the truncated polynomials $A_m(w^{-1}, z^{-1})$, $B_m(w^{-1}, z^{-1})$ and $C_m(w^{-1}, z^{-1})$ near the edges share the same parameters as their counterparts away from the edges seems unlikely to be valid in most cases. A more general model would be to allow these polynomials to have different parameter values for very small or very large m.

This is more easily described in terms of the multivariable description. In §3.1.1 we had for example (Equation 3.3.1)

$$\underline{\underline{A}}(z^{-1}) = \begin{bmatrix} a'_0(z^{-1}) & a'_{-1}(z^{-1}) & \cdots & & & 0 \\ a'_1(z^{-1}) & a'_0(z^{-1}) & \ddots & & & \\ \vdots & \ddots & \ddots & \ddots & & \vdots \\ & & \ddots & a'_0(z^{-1}) & a'_{-1}(z^{-1}) \\ 0 & & \cdots & a'_1(z^{-1}) & a'_0(z^{-1}) \end{bmatrix} \quad (3.3.12)$$

where the elements of each diagonal are the same. Recalling that the m'th row of $\underline{\underline{A}}(z^{-1})$ corresponds to $A_m(w^{-1}, z^{-1})$ our new edge conditions can be incorporated by a new $\underline{\underline{A}}(z^{-1})$ given by

$$\underline{\underline{A}} = \begin{bmatrix} a''_{1,0} & a''_{1,-1} & \cdots & & & & & 0 \\ \vdots & \ddots & & & & & & \\ a''_{m_l,1} & a''_{m_l,0} & a''_{m_l,-1} & & & & & \\ & a'_1 & a'_0 & a'_{-1} & & & & \\ & & \ddots & \ddots & \ddots & & & \\ & & & a'_1 & a'_0 & a'_{-1} & & \\ & & & & a''_{m_r,1} & a''_{m_r,0} & a''_{m_r,-1} & \\ & & & & & & \ddots & \vdots \\ 0 & & & & & \cdots & a''_{W,1} & a''_{W,0} \end{bmatrix}$$

Note that away from the top and bottom (corresponding to distance from the edges) the matrix retains the neat symmetric structure of Equation 3.3.12. Furthermore the sparseness of the structure is preserved.

If we know how the parameters behave we can still construct our predictor and controllers algorithmically as in §3.1, provided we modify our substitutions for future values of y suitably near the edges.. Furthermore many of our results continue to hold given certain modifications. Suppose we modify $A_m(w^{-1}, z^{-1})$ only when $m \leq m_l$ and $m \geq m_r$ for some m_l and m_r. Then for example Theorem 3.1.2 still remains true except that the condition expressed in Equation 3.1.7 becomes

$$1 + k_2 M + m_l \leq m + k_1 \leq m_r - k_2 M - 1$$

In §4.5.3 (below) we will show how to construct a parameter estimation routine which takes into account such behaviour at the edges and we will show some simulation results of a self-tuning controller for such a model in §5.1.3 (below).

3.4 Other causality structures

3.4.1 The process

So far for the two-dimensional control problem we have considered models of the form

$$A(w^{-1},z^{-1})y(m,n) = z^{-\nu}B(w^{-1},z^{-1})u(m,n) + C(w^{-1},z^{-1})e(m,n)$$

where $A(w^{-1},z^{-1})$, $B(w^{-1},z^{-1})$ and $C(w^{-1},z^{-1})$ all have NSHP (non-symmetric half-plane) local supports. That is to say

$$A(w^{-1},z^{-1}) = \sum_{i=0}^{M}\sum_{j=0}^{N} a_{i,j} w^{-i} z^{-j} + \sum_{i=1}^{M}\sum_{j=1}^{N} a_{-i,j} w^{i} z^{-j}$$

and similarly for $B(w^{-1},z^{-1})$ and $C(w^{-1},z^{-1})$.

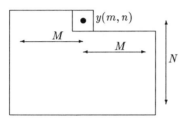

Fig 3.4.1 *NSHP (non-symmetric half-plane) local support.*

The choice of such local supports was made to correspond with [Wagner 1987] and [Caldas-Pinto 1983] where two-dimensional ARMA models are considered (ie there is no B polynomial). Here the choice of NSHP local supports could be justified on two counts. From a practical point of view it allowed a raster scan and hence a sensible recursion path through the data. From a theoretical point of view the NSHP *global* support is the natural choice for two-dimensional stochastic processes [Whittle 1954], [O'Connor and Huang 1981].

For a two-dimensional CARMA process the NSHP global support no longer makes such good physical sense. For a process in which one dimension is temporal (with strict causality) and the other spatial (with no causality) the natural environment is the SHP (symmetric half-plane). Thus the ideal model would be

$$A(w^{-1},z^{-1})y(m,n) = z^{-\nu}B(w^{-1},z^{-1})u(m,n) + C(w^{-1},z^{-1})e(m,n)$$

where $A(w^{-1}, z^{-1})$, $B(w^{-1}, z^{-1})$ and $C(w^{-1}, z^{-1})$ all have SHP (symmetric half-plane) local supports; for example

$$A(w^{-1}, z^{-1}) = \sum_{i=0}^{M}\sum_{j=0}^{N} a_{i,j} w^{-i} z^{-j} + \sum_{i=1}^{M}\sum_{j=0}^{N} a_{-i,j} w^{i} z^{-j}$$

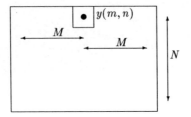

Fig 3.4.2 *SHP (symmetric half-plane) local support.*

Computationally such a local support would be difficult to handle in the two-dimensional context considered here (note such a support is valid if we model the process as multivariable as in §3.3, but we have the usual problem that the number of variables becomes prohibitive). In the simulations of §3.2 to avoid computational difficulties we preferred to set

$$A(w^{-1}, 0) = 1$$

We saw in §3.1 that the computational advantages in this case are firstly that the partition

$$C = AF + z^{-\nu}G$$

generates finite-order polynomials $F(w^{-1}, z^{-1})$ and $G(w^{-1}, z^{-1})$ directly and secondly that the edges may be ignored when the current pixel is sufficiently far from the edges. We will call this third case the SWP (symmetric wedged-plane) support.

In §3.4.3 (below) we will derive a new controller for the hybrid model

$$A(w^{-1}, z^{-1})y(m, n) = z^{-\nu} B(w^{-1}, z^{-1})u(m, n) + C(w^{-1}, z^{-1})e(m, n)$$

where $A(w^{-1}, z^{-1})$ has SWP support and $C(w^{-1}, z^{-1})$ has NSHP support as usual but $B(w^{-1}, z^{-1})$ has SHP support. In §3.4.2 (below) we will see that this allows richer

3.4 Other causality structures

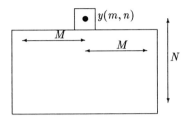

Fig 3.4.3 *SWP (symmetric wedge-plane) local support.*

dynamics to be modelled than our usual model. Strictly we may allow $A(w^{-1}, z^{-1})$ to have NSHP support but it is assumed that in practice the computational benefit of choosing $A(w^{-1}, z^{-1})$ to have SWP support would be over-riding.

3.4.2 Model responses to step inputs

In this section we present the response of various models to a step input in just one column. We will not consider the effect of the structure of $C(w^{-1}, z^{-1})$; in fact to clarify results we will consider only deterministic models of the form

$$A(w^{-1}, z^{-1}) y(m, n) = z^{-\nu} B(w^{-1}, z^{-1}) u(m, n)$$

In particular we will consider the nine different models given by

$$A_i(w^{-1}, z^{-1}) y(m, n) = z^{-2} B_j(w^{-1}, z^{-1}) u(m, n) \text{ for } 1 \leq i \leq 3 \text{ and } 1 \leq j \leq 3$$

where

$$A_1(w^{-1}, z^{-1}) = 1 - a_1 w^{-1} - a_2 z^{-1} - a_1 w$$

$$A_2(w^{-1}, z^{-1}) = 1 - a_1 w^{-1} - a_2 z^{-1} - a_1 w z^{-1}$$

$$A_3(w^{-1}, z^{-1}) = 1 - a_1 w^{-1} z^{-1} - a_2 z^{-1} - a_1 w z^{-1}$$

and

$$B_1(w^{-1}, z^{-1}) = b_0 + b_1 w^{-1} + b_1 w$$

$$B_2(w^{-1}, z^{-1}) = b_0 + b_1 w^{-1} + b_1 w z^{-1}$$
$$B_3(w^{-1}, z^{-1}) = b_0 + b_1 w^{-1} + b_1 w z^{-1}$$

Note that A_1 and B_1 are SHP causal, A_2 and B_2 are NSHP causal and A_3 and B_3 are SWP causal.

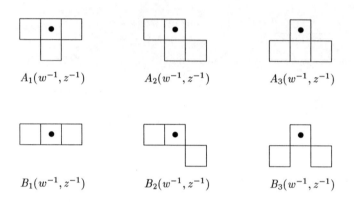

Fig 3.4.4 *Local supports.*

Note also that when
$$A(w^{-1}, z^{-1}) = A_1(w^{-1}, z^{-1})$$
then the value of $y(m, n)$ cannot be calculated in the conventional manner; instead a whole row of y's must be evaluated at once as the solution of a set of simultaneous equations. (This work is partly inspired by finite-difference or net methods for the approximation of partial differential equations—see for example [Richtmyer 1957]. There seems to be a direct correspondence between on the one hand NSHP causal processes and explicit net methods and on the other hand SHP causal processes and implicit net methods.)

The following values for the a_i's and b_i's were chosen arbitrarily:

$a_1 = 0.35$, $a_2 = 0.2$, $b_0 = 2$ and $b_1 = 0.9$

In each case $y(m, n)$ was calculated over 30×30 pixels with $u(m, n)$ zero everywhere except

$m = 15$ and $n \geq 10$

3.4 Other causality structures

where

$$u(m,n) = 1$$

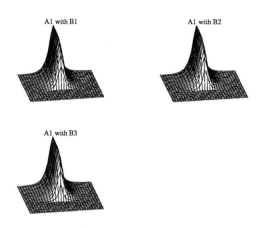

Fig 3.4.5 *Step responses with $A(w^{-1}, z^{-1})$ SHP-causal.*

In Fig 3.4.5 we show the two-dimensional 'mesh' of the output in the three cases where $A(w^{-1}, z^{-1})$ is SHP-causal. In Fig 3.4.6 we show the corresponding initial responses—ie the first row where the output is non-zero. Similar results are shown in Figs 3.4.7 and 3.4.8 for when $A(w^{-1}, z^{-1})$ is NSHP-causal and in Figs 3.4.9 and 3.4.10 for when $A(w^{-1}, z^{-1})$ is SWP-causal. From the results we can make the following observations. Firstly all the models have the same steady-state response. This is intuitively obvious since if $y_{ss}(m)$ and $u_{ss}(m)$ are respectively the steady-state outputs and inputs then

$$y_{ss}(m) = \frac{B(w^{-1}, 1)}{A(w^{-1}, 1)} u_{ss}(m)$$

which is equal in each case. Therefore the choice of $A_i(w^{-1}, z^{-1})$ and $B_j(w^{-1}, z^{-1})$ affects only the initial dynamic response. Note that in the models currently used for cross-directional control on paper machines only the steady state is considered [Wilhelm and Fjeld 1983]. The second observation is that the initial responses are all very different.

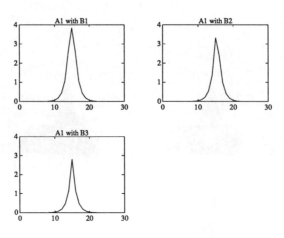

Fig 3.4.6 *Initial responses to step input with $A(w^{-1}, z^{-1})$ SHP-causal.*

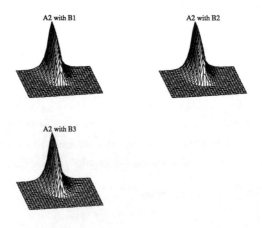

Fig 3.4.7 *Step responses with $A(w^{-1}, z^{-1})$ NSHP-causal.*

3.4 Other causality structures

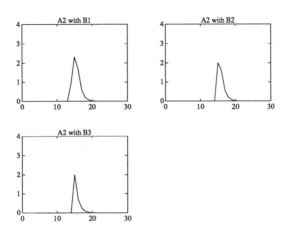

Fig 3.4.8 *Initial responses to step input with $A(w^{-1}, z^{-1})$ NSHP-causal.*

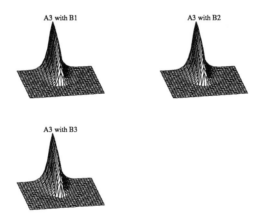

Fig 3.4.9 *Step responses with $A(w^{-1}, z^{-1})$ SWP-causal.*

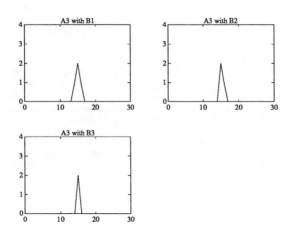

Fig 3.4.10 *Initial responses to step input with $A(w^{-1}, z^{-1})$ SWP-causal.*

Heuristically the step response to a single actuator change on a paper machine is as follows. Initially there is a large delay due to the distance between the lip and scanning gauge. Furthermore the time response of the actuator itself may be large, especially if the screw is of the thermally heated type. However the change in the lip profile will be relatively instantaneous across a fairly wide cross-section (Fig 3.4.11). Thereafter there will be a slow dynamic response as the effects of this change spread across the web itself.

From the simulations it is clear that choosing $A(w^{-1}, z^{-1})$ to be SHP-causal represents this kind of behaviour best. In this case the 'initial' response is spread over a width of at least 10 visible pixels. With $A(w^{-1}, z^{-1})$ NSHP-causal there is a similar broad initial response to one side only. However given that such choices for $A(w^{-1}, z^{-1})$ are unreasonable and that $A(w^{-1}, z^{-1})$ is constrained to be SWP-causal then choosing $B(w^{-1}, z^{-1})$ to be SHP-causal offers a reasonable representation of this kind of behaviour. It can be seen that the initial response is spread over 3 pixels in this case (corresponding to the dimensions of $B(w^{-1}, z^{-1})$ itself) as opposed to 2 when $B(w^{-1}, z^{-1})$ is NSHP-causal and 1 when $B(w^{-1}, z^{-1})$ is SWP-causal This is our justification for considering this kind of model.

3.4 Other causality structures

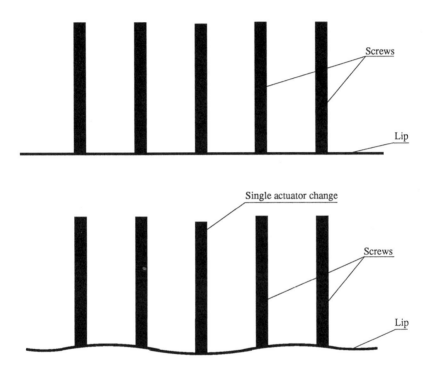

Fig 3.4.11 *A single actuator (screw) change causing the lip profile to change over a wide area.*

3.4.3 Control—an informal approach

We are now concerned with the general model

$$A(w^{-1}, z^{-1})y(m,n) = z^{-\nu}B(w^{-1}, z^{-1})u(m,n) + C(w^{-1}, z^{-1})e(m,n)$$

where $A(w^{-1}, z^{-1})$ is SWP-causal, $B(w^{-1}, z^{-1})$ is SHP-causal and finally $C(w^{-1}, z^{-1})$ is NSHP-causal. We will develop a generalised minimum variance type controller corresponding to that developed in §3.1.4.

Note that it is possible to express the process as causal if we allow a horizontal component in the delay (cf §2.1). Specifically we could put

$$A(w^{-1}, z^{-1})y(m,n) = w^M z^{-\nu} B'(w^{-1}, z^{-1}) + C(w^{-1}, z^{-1})e(m,n) \qquad (3.4.1)$$

where

$$B'(w^{-1}, z^{-1}) = w^{-\mu} B(w^{-1}, z^{-1}) \qquad (3.4.2)$$

and is NSHP-causal. However in constructing a controller for such a process we would find no controlling inputs for the last M outputs at the righthand side of the web (see Fig 3.4.12). Such a modification of the model does not then offer a shortcut to the derivation of a controller.

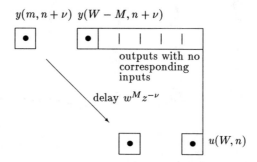

Fig 3.4.12 *Relation of outputs to inputs at the righthand edge for the process modelled by Equations 3.4.1 and 3.4.2.*

Instead suppose we construct the usual pseudo-output

$$\begin{aligned}\phi(m,n) &= P(w^{-1}, z^{-1})y(m,n) + z^{-\nu}Q(w^{-1}, z^{-1})u(m,n) \\ &\quad - z^{-\nu}R(w^{-1}, z^{-1})r(m,n)\end{aligned}$$

3.4 Other causality structures

(We may assume that as usual $P(w^{-1}, z^{-1})$, $Q(w^{-1}, z^{-1})$ and $R(w^{-1}, z^{-1})$ are all SWP-causal though in fact we may generalise to $Q(w^{-1}, z^{-1})$ and $R(w^{-1}, z^{-1})$ being SHP-causal.) Then as usual we may substitute for $y(m, n)$ to obtain

$$\begin{aligned}\phi(m, n+\nu) &= X_m(w^{-1}, z^{-1})y(m,n) + Y_m(w^{-1}, z^{-1})u(m,n) \\ &\quad + Z_m(w^{-1}, z^{-1})e(m,n) - R(w^{-1}, z^{-1})r(m,n) \\ &\quad + \{\text{future noise}\}\end{aligned}$$

Here $X_m(w^{-1}, z^{-1})$ and $Z_m(w^{-1}, z^{-1})$ are NSHP-causal while $Y_m(w^{-1}, z^{-1})$ is SHP-causal. In fact it is easy to see that if

$$P(0,0) = p_{0,0}$$

$$B(w^{-1}, 0) = \sum_{i=-M}^{M} b_{i,0} w^{-i}$$

and

$$Q(w^{-1}, 0) = \sum_{i=-M}^{M} q_{i,0} w^{-i}$$

then

$$Y_m(w^{-1}, 0) = \sum_{i=-M}^{M} (p_{0,0} b_{i,0} + q_{i,0}) w^{-i}$$

As usual we may construct recursively an estimate $\hat{e}(m, n)$ of the noise using the model:

$$C(w^{-1}, z^{-1})\hat{e}(m,n) = A(w^{-1}, z^{-1})y(m,n) - z^{-\nu}B(w^{-1}, z^{-1})u(m,n)$$

Then we would like to implement the controller

$$\begin{aligned}X_m(w^{-1}, z^{-1})y(m,n) + Y_m(w^{-1}, z^{-1})u(m,n) + Z_m(w^{-1}, z^{-1})\hat{e}(m,n) \\ = R(w^{-1}, z^{-1})r(m,n)\end{aligned}$$

We may represent this as

$$\sum_{i=-M}^{M} (p_{0,0} b_{i,0} + q_{i,0}) w^{-i} u(m,n) + \{\text{known values}\} = 0$$

where at (m, n) we have a numerical value for the {known values}.

Of course this represents a non-causal (in the w direction) control law and we cannot implement it as such. This non-causality stems directly from our choice of the support for $B(w^{-1}, z^{-1})$. Suppose however we *delay* implementing the controller until the end of the scan (which we assume runs from $m = 1$ to $m = W$). We will then have W such control laws which we may represent in multivariable terms as

$$\underline{\underline{Y}}\,\underline{u}(n) = \underline{d}$$

Here $\underline{\underline{Y}}$ is a known $W \times W$ matrix, $\underline{u}(n)$ a $W \times 1$ column vector whose m'th entry is $u(m, n)$ and $\underline{d}$ a $W \times 1$ column vector with known entries. Our control is then

$$\underline{u}(n) = \underline{\underline{Y}}^{-1}\underline{d}$$

Note that $\underline{\underline{Y}}$ is non-dynamic; that is to say it has no z^{-1} elements—its entries are all numerical values. Furthermore it is sparse; schematically we may represent its structure as

$$\underline{\underline{Y}} = \begin{bmatrix} \ddots & \ddots & & 0 \\ \ddots & \ddots & \ddots & \\ & \ddots & \ddots & \ddots \\ 0 & & \ddots & \ddots \end{bmatrix}$$

This allows us to perform Gaussian elimination in a particularly efficient recursive manner.

Specifically we can borrow from finite-difference approximation methods [Richtmyer 1957] the same technique mentioned in §3.4.2 in the context of simulating $y(m, n)$ when $A(w^{-1}, z^{-1})$ was SHP-causal. We will consider it in some detail here. To repeat, our control law at each (m, n) is

$$\sum_{i=-M}^{M} (p_{0,0} b_{i,0} + q_{i,0})\, w^{-i} u(m, n) = k_m \tag{3.4.3}$$

for some known k_m. We would like to reformulate this as

$$u(m, n) = \sum_{j=1}^{M} \lambda_{j,m} u(m+j, n) + \lambda_{M+1,m} \tag{3.4.4}$$

for some coefficients λ. We may obtain such an expression recursively as follows:

3.4 Other causality structures

Suppose we already know the values of the λ's in the expressions

$$u(m-i,n) = \sum_{j=1}^{M} \lambda_{j,m-i} u(m-i+j,n) + \lambda_{M+1,m-i} \qquad (3.4.5)$$

for all $i \geq 1$. Then we can substitute for $u(m-i,n)$ in the control law given by Equation 3.4.3 to obtain the new control law

$$u(m,n) = \sum_{j=1}^{M} \lambda_{j,m} u(m+j,n) + \lambda_{M+1,m} \qquad (3.4.6)$$

as desired.

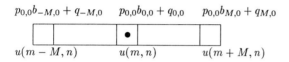

Fig 3.4.13 *Pictorial representation of Equation 3.4.3.*

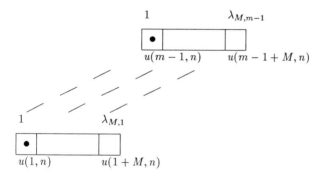

Fig 3.4.14 *Pictorial representation of Equations 3.4.5.*

Hence we can obtain such an expression for all m recursively in order of *increasing* m; then in order of decreasing m we can recursively calculate the actual control law. We will illustrate this with an example below.

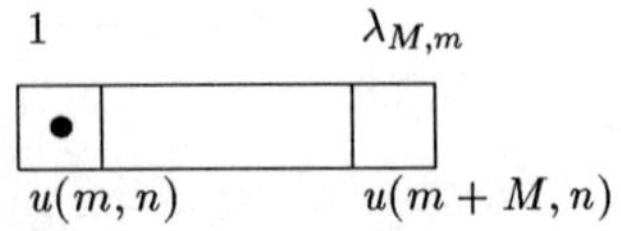

Fig 3.4.15 *Pictorial representation of Equation 3.4.6 formed by substituting from Equations 3.4.4 successively into Equation 3.4.3.*

3.4.4 A simulation example

Suppose we have the process

$$\left(1 - a_1 w^{-1} z^{-1} - a_2 z^{-1} - a_1 w z^{-1}\right) y(m,n) = z^{-1}\left(b_1 w^{-1} + b_0 + b_1 w\right) u(m,n) + e(m,n)$$

and we wish to implement integral control. The most simple choice of pseudo-output would be

$$\phi(m,n) = y(m,n) + z^{-1}\left(1 - z^{-1}\right) u(m,n) - z^{-1} r$$

where r is our chosen setpoint. It is easy to see that

$$\begin{aligned}\phi(m, n+1) &= b_1 u(m-1, n) + (1 + b_0) u(m, n) + b_1 u(m+1, n) \\ &\quad + (a_1 b_1 - 1) u(m, n-1) + a_1 b_0 u(m+1, n-1) \\ &\quad + a_1 b_1 u(m+2, n-1) \\ &\quad + a_1 y(m-1, n) + a_2 y(m, n) \\ &\quad + a_1^2 y(m, n-1) + a_1 a_2 y(m+1, n-1) + a_1^2 y(m+2, n-1) \\ &\quad - r + e(m, n+1) + a_1 e(m+1, n)\end{aligned}$$

(with appropriate modifications at the edges).

Then the control law is

$$b_1 u(m-1, n) + (1 + b_0) u(m, n) + b_1 u(m+1, n) = k_m$$

where

$$\begin{aligned}k_m &= r + (1 - a_1 b_1) u(m, n-1) - a_1 b_0 u(m+1, n-1) - a_1 b_1 u(m+2, n-1) \\ &\quad - a_1 y(m-1, n) - a_2 y(m, n) \\ &\quad - a_1^2 y(m, n-1) - a_1 a_2 y(m+1, n-1) - a_1^2 y(m+2, n-1)\end{aligned}$$

3.4 Other causality structures

This control law is non-causal, but if we desire a control law of the form

$$u(m,n) = \lambda_{1,m} u(m+1,n) + \lambda_{2,m}$$

then we have the recursive relationships:

$$\lambda_{1,1} = \frac{-b_1}{1+b_0}$$

$$\lambda_{2,1} = \frac{k_1}{1+b_0}$$

$$\lambda_{1,m} = \frac{-b_1}{1+b_0+b_1\lambda_{1,m-1}}$$

$$\lambda_{2,m} = \frac{k_m - b_1\lambda_{2,m-1}}{1+b_0+b_1\lambda_{1,m-1}}$$

Thus as we receive data we may calculate the values of the λ's. Then at the end of each scan we may calculate the values of the $u(m,n)$'s for all m. Here we show the output and input for such a controller generated over 25×25 pixels. The parameters were chosen to be the same as in §3.4.2, and the setpoint was chosen to be 100. We show a mesh of the output together with a cross-section of the final row to give a sense of scale.

3.4.5 Control—a more formal approach

In §3.4.3 we constructed a new two-dimensional controller for a new two-dimensional control problem. In §3.4.4 we showed with a simple example that it can provide an effective stable control strategy, but so far we have provided no theoretical justification for the controller. In particular we have allowed ourselves to delay implementing the control action $u(m,n)$ until the end of the current scan; this tampering with the 'rules of the game' needs some careful analysis and justification.

As mentioned above we have altered the class of admissible controllers. For our usual two-dimensional control problem we defined (§2.4) an admissible controller to be one where $u(m,n)$ is a function of

$$y(m-i,n-j) \; \forall (m-i,n-j) \preceq (m,n)$$

and

$$u(m-i,n-j) \; \forall (m-i,n-j) \prec (m,n)$$

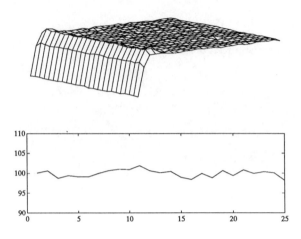

Fig 3.4.16 *Closed-loop output when $B(w^{-1}, z^{-1})$ has SHP support.*

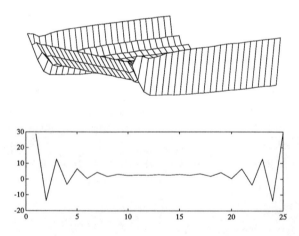

Fig 3.4.17 *Input when $B(w^{-1}, z^{-1})$ has SHP support.*

3.4 Other causality structures

However here we have allowed $u(m, n)$ to be a function of $u(m + i, n)$ for $i \geq 1$ as well. An admissible controller is then one for which $u(m, n)$ is a function of

$$y(m - i, n - j) \; \forall (m - i, n - j) \preceq (W, n)$$

and

$$u(m - i, n - j) \; \forall (m - i, n - j) \preceq (W, n)$$

At first sight it may seem that we are getting something for nothing here as we are assuming more information is available when $u(m, n)$ is calculated. In practice of course we are at the same time delaying the implementation of $u(m, n)$, so the delay in our original model would be greater. On the other hand the new strategy has the advantage that it is in line with current practice in papermaking and we are no longer constrained to synchronise our actuator signals with the position of the scan. It also means that we can have more flexibility with any additional constraints or jacketing software on the control action. For example if we require the cross-sectional area of the lip to be constant it is relatively simple to scale the total row of inputs appropriately at the end of each scan.

Our new controller is not optimal in the sense that we do not achieve the minimum variance control for the pseudo-output ϕ. Although we have allowed $u(m, n)$ to be a function of $y(m + i, n)$ for $i \geq 1$ we have chosen not to use these values explicitly for computational reasons. In terms of partition (if we ignore the effects of the edges for the moment) we have closed-loop output

$$\phi(m, n) = F_1(w^{-1}, z^{-1}) e(m, n)$$

with $F_1(w^{-1}, z^{-1})$ given by

$$C = AF_1 + z^{-\nu} G_1$$

whereas the optimal output would be

$$\phi(m, n) = F_2(w^{-1}, z^{-1}) e(m, n)$$

with $F_2(w^{-1}, z^{-1})$ given by

$$C = AF_2 + w^{W-m} z^{-\nu} G_2$$

and

$$E\left[F_2(w^{-1}, z^{-1}) e(m, n)\right]^2 \leq E\left[F_1(w^{-1}, z^{-1}) e(m, n)\right]^2$$

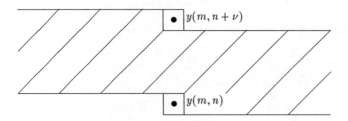

Fig 3.4.18 Support for $F_1(w^{-1}, z^{-1})$.

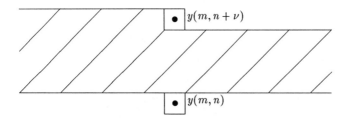

Fig 3.4.19 Support for $F_2(w^{-1}, z^{-1})$.

Note that $F_1(w^{-1}, z^{-1})$ and $F_2(w^{-1}, z^{-1})$ are very similar polynomials. If we define

$$H(w^{-1}, z^{-1}) = \frac{C(w^{-1}, z^{-1})}{A(w^{-1}, z^{-1})}$$

$$= \sum_{i=0}^{\infty} h_{i,0} w^{-i} + \sum_{j=1}^{\infty} \sum_{i=-jM}^{\infty} h_{i,j} w^{-i} z^{-j}$$

then

$$F_1(w^{-1}, z^{-1}) = \sum_{i=0}^{\infty} h_{i,0} w^{-i} + \sum_{j=1}^{\nu-1} \sum_{i=-jM}^{\infty} h_{i,j} w^{-i} z^{-j} + \sum_{i=-\nu M}^{-1} h_{i,\nu} w^{-i} z^{-\nu}$$

and

$$F_2(w^{-1}, z^{-1}) = \sum_{i=0}^{\infty} h_{i,0} w^{-i} + \sum_{j=1}^{\nu-1} \sum_{i=-jM}^{\infty} h_{i,j} w^{-i} z^{-j}$$

3.4 Other causality structures

so that

$$F_1(w^{-1}, z^{-1}) - F_2(w^{-1}, z^{-1}) = \sum_{i=-\nu M}^{-1} h_{i,\nu} w^{-i} z^{-\nu}$$

Hence the difference is a finite sum of small terms so that in most cases the quantity

$$E\left[F_1(w^{-1}, z^{-1})e(m,n)\right]^2 - E\left[F_2(w^{-1}, z^{-1})e(m,n)\right]^2 = \sigma^2 \sum_{i=-\infty}^{-1} h_{i,\nu}^2$$

will be small. In other words our controller, though sub-optimal, is close to the optimal.

The discussion is best illustrated in terms of the example of §3.4.4. Recall that we expressed ϕ as

$$\begin{aligned}\phi(m, n+1) &= b_1 u(m-1, n) + (1+b_0)u(m,n) + b_1 u(m+1, n) \\ &+ (a_1 b_1 - 1)u(m, n-1) + a_1 b_0 u(m+1, n-1) \\ &+ a_1 b_1 u(m+2, n-1) \\ &+ a_1 y(m-1, n) + a_2 y(m, n) \\ &+ a_1^2 y(m, n-1) + a_1 a_2 y(m+1, n-1) + a_1^2 y(m+2, n-1) \\ &- r + e(m, n+1) + a_1 e(m+1, n)\end{aligned}$$

(with appropriate modifications at the edges). Then the control law was

$$b_1 u(m-1, n) + (1+b_0)u(m,n) + b_1 u(m+1, n) = k_m$$

where

$$\begin{aligned}k_m &= r + (1 - a_1 b_1)u(m, n-1) - a_1 b_0 u(m+1, n-1) - a_1 b_1 u(m+2, n-1) \\ &- a_1 y(m-1, n) - a_2 y(m, n) \\ &- a_1^2 y(m, n-1) - a_1 a_2 y(m+1, n-1) - a_1^2 y(m+2, n-1)\end{aligned}$$

For the optimal control we re-express ϕ as

$$\begin{aligned}\phi(m, n+1) &= b_1 u(m-1, n) + (1+b_0)u(m,n) + b_1 u(m+1, n) \\ &- u(m, n-1) \\ &+ a_1 y(m-1, n) + a_2 y(m, n) + a_1 y(m+1, n) \\ &- r + e(m, n+1)\end{aligned}$$

(with appropriate modifications at the edges). Then the control law is

$$b_1 u(m-1, n) + (1+b_0)u(m,n) + b_1 u(m+1, n) = k'_m - a_1 y(m+1, n)$$

where

$$k'_m = r + u(m, n-1) - a_1 y(m-1, n) - a_2 y(m, n)$$

Suppose we wish now to have a controller of the form

$$u(m, n) = \lambda'_{1,m} u(m+1, n) + \lambda'_{2,m} + \lambda'_{3,m} y(m+1, n)$$

Then we have the recursive relations

$$\lambda'_{1,1} = \frac{-b_1}{1 + b_0}$$

$$\lambda'_{2,1} = \frac{k'_1}{1 + b_0}$$

$$\lambda'_{3,1} = \frac{-a_1}{1 + b_0}$$

$$\lambda'_{1,m} = \frac{-b_1}{1 + b_0 + b_1 \lambda'_{1,m-1}}$$

$$\lambda'_{2,m} = \frac{k'_m - b_1 \lambda'_{2,m-1} - b_1 \lambda'_{3,m-1} y(m, n)}{1 + b_0 + b_1 \lambda'_{1,m-1}}$$

$$\lambda'_{3,m} = \frac{-a_1}{1 + b_0 + b_1 \lambda'_{1,m-1}}$$

In Figs 3.4.20 and 3.4.21 we show the output and input respectively for such a controller. A slight difference is discernible from the results in §3.4.4 (remember the 'optimality' is in the variance of the pseudo-output which is not shown).

In some cases it may be desirable to implement such a controller. However in general there are two good reasons for not doing so. Firstly the 'optimal' controller will usually involve more computation. In the example it can be seen we have a third set of λ's to calculate—the $\lambda'_{3,m}$'s. In the example this is set against a more simple partition, but in general, and in particular when delays are large, there will be many more such sets of λ's making the computational cost prohibitive. Secondly for the 'optimal' control we lose the neat two-dimensional algebraic properties of the partition; this is indicated above. In fact the partition for the 'optimal' control is that for the conventional multivariable problem, without the modification introduced in §3.3.2.

Note that we have not formulated a cost function so our philosophy has strayed from that in §3.1. Instead we repeat that the controller is close to the optimal (in

3.4 Other causality structures

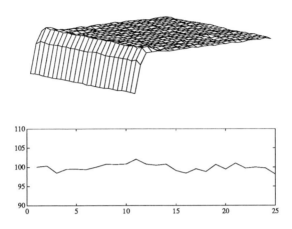

Fig 3.4.20 *Closed-loop output when $B(w^{-1}, z^{-1})$ has SHP support using the 'optimal' controller.*

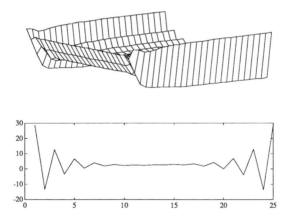

Fig 3.4.21 *Input when $B(w^{-1}, z^{-1})$ has SHP support using the 'optimal' controller.*

the conventional sense) but retains the neat properties of two-dimensional control. In particular there are two important points. Firstly the pseudo-output is an MA (moving average) process. Secondly, the partition remains the two-dimensional partition, and so the two-dimensional algebraic analysis of §2 remains valid given the constraints discussed in §3.1 and §3.3.

For example it is easy to see that (if we allow for the edges) we are effectively performing the partition

$$PC = AF' + z^{-\nu}G'$$

Since we assume that $A(w^{-1}, z^{-1})$ is SWP-causal we have

$$A(w^{-1}, 0) = 1$$

and so both $F'(w^{-1}, z^{-1})$ and $G'(w^{-1}, z^{-1})$ are finite-order polynomials (§2.4.2). The control law can then be characterised as

$$(BF' + QC)\,u(m,n) + G'y(m,n) = RCr$$

(N.B. this is a non-causal control law). The closed-loop output is then

$$y(m,n) = \frac{z^{-\nu}BR}{PB + AQ}r + \frac{BF' + QC}{PB + AQ}e(m,n)$$

This result is exactly the same as for the case with $B(w^{-1}, z^{-1})$ NSHP-causal.

4 Parameter estimation

In this chapter we consider parameter estimation for two-dimensional processes. The discussion will be independent of any control or prediction algorithm and so we will confine ourselves here to ARMA processes. We begin in §4.1 by examining the generic recursive least squares (RLS), recursive extended least squares (RELS) and approximate maximum likelihood (AML) algorithms in two dimensions. In §4.2 we examine the forgetting strategy used in [Caldas-Pinto 1983] and [Wagner 1987] and consider a more suitable criterion for weighting data in two dimensions. In §4.3 and §4.4 we consider aspects of the algorithmic implementation of such a forgetting strategy; in §4.3 we examine the order in which we recurse through the data while in §4.4 we consider the use of QR factorisation of the information matrix involved in the forgetting. In §4.5 we consider two special cases which allow a more simple updating scheme; a modified form of the second allows us to deal in a neat way with varying parameters at the edge of the data field as considered in §3.3.7.

4.1 Least squares estimation in two dimensions

4.1.1 AR processes

In both [Caldas-Pinto 1983] and [Wagner 1987] a novel estimation algorithm is used to identify the parameters of the two-dimensional polynomials $A(w^{-1}, z^{-1})$ and $C(w^{-1}, z^{-1})$ in a two-dimensional ARMA model. This algorithm is based on the RELS (recursive extended least squares) algorithm for one-dimensional polynomials (see for example [Ljung and Soderstrom 1983] or [Wellstead and Zarrop 1991]). In many respects the two-dimensional recursive formulation is strikingly similar to the one-dimensional. However in neither [Caldas-Pinto 1983] nor [Wagner 1987] is the new algorithm (which we will term 2D-RELS from now on) either derived or stated. We will here derive the equivalent estimator for the special case of a two-dimensional AR model (where $C(w^{-1}, z^{-1}) = 1$); following from the one-dimensional nomenclature we will term this estimator 2D-RLS. The extension of this to 2D-RELS and the closely-related 2D-AML (two-dimensional approximate maximum likelihood) is discussed in §4.1.2.

Suppose we wish to identify the parameters of the two-dimensional AR model

$$A(w^{-1}, z^{-1})y(m,n) = e(m,n)$$

where

$$A(w^{-1}, z^{-1}) = \sum_{i=0}^{M_a}\sum_{j=0}^{N_a} a_{i,j} w^{-i} z^{-j} + \sum_{i=1}^{M_a}\sum_{j=1}^{N_a} a_{-i,j} w^{i} z^{-j}$$

Define the true parameter vector

$$\theta^T = (-a_{1,0}, ..., -a_{M_a,0}, -a_{-M_a,1}, ..., -a_{M_a,1}, ..., -a_{-M_a,N_a}, ..., -a_{M_a,N_a})$$

Define also the parameter estimator vector

$$\hat{\theta}^T(m,n) = (..., -\hat{a}_{i,j}, ...)$$

and the corresponding data vector

$$\phi^T(m,n) = (..., y(m-i, n-j), ...)$$

Thus we may write the AR model as

$$y(m,n) = \phi^T(m,n)\theta + e(m,n)$$

If we form the residuals

$$\eta(m-i, n-j) = y(m-i, n-j) - \phi^T(m-i, n-j)\hat{\theta}(m,n)$$

then suppose we wish to minimise the cost function

$$J(m,n) = \sum_i \sum_j \eta(m-i, n-j)^2$$

where the summation is over the global non-symmetric half-plane past.

We have

$$\frac{\partial J}{\partial \hat{\theta}} = -2 \sum_i \sum_j \left(y(m-i, n-j) - \phi^T(m-i, n-j)\hat{\theta}(m,n) \right) \phi(m-i, n-j)$$

and

$$\frac{\partial^2 J}{\partial \hat{\theta}^2} = 2 \sum_i \sum_j \phi(m-i, n-j)\phi^T(m-i, n-j)$$

$$> 0$$

4.1 Least squares estimation in two dimensions

so we have a minimum or least squares solution when

$$\hat{\theta}(m,n) = \left[\sum_i\sum_j \phi(m-i,n-j)\phi^T(m-i,n-j)\right]^{-1}$$
$$\times \left[\sum_i\sum_j \phi(m-i,n-j)y(m-i,n-j)\right]$$

For a recursive algorithm we assume that a raster scan of the data is used. We have the relations

$$\sum_i\sum_j \phi(m-i,n-j)y(m-i,n-j) =$$

$$\begin{cases} \sum_i\sum_j \phi(m-1-i,n-j)y(m-1-i,n-j) \\ + \phi(m,n)y(m,n) & \text{for } m > 1 \\[2ex] \sum_i\sum_j \phi(W-i,n-1-j)y(W-i,n-1-j) \\ + \phi(1,n)y(1,n) & \text{for } m = 1 \end{cases}$$

and

$$\sum_i\sum_j \phi(m-i,n-j)\phi^T(m-i,n-j) =$$

$$\begin{cases} \sum_i\sum_j \phi(m-1-i,n-j)\phi^T(m-1-i,n-j) \\ + \phi(m,n)\phi^T(m,n) & \text{for } m > 1 \\[2ex] \sum_i\sum_j \phi(W-i,n-1-j)\phi^T(W-i,n-1-j) \\ + \phi(1,n)\phi^T(1,n) & \text{for } m = 1 \end{cases}$$

Furthermore we may invoke the matrix inversion lemma [Ljung and Soderstrom 1983] in the same way as in the one-dimensional case: suppose

$$P(m,n) = \left[\sum_i\sum_j \phi(m-i,n-j)\phi^T(m-i,n-j)\right]^{-1}$$

Then

$$P(m,n) = \begin{cases} P(m-1,n)\left[I - \dfrac{\phi(m,n)\phi^T(m,n)P(m-1,n)}{1+\phi^T(m,n)P(m-1,n)\phi(m,n)}\right] & \text{for } m > 1 \\[2ex] P(W,n-1)\left[I - \dfrac{\phi(1,n)\phi^T(1,n)P(W,n-1)}{1+\phi^T(1,n)P(W,n-1)\phi(1,n)}\right] & \text{for } m = 1 \end{cases}$$
(4.1.1)

Hence going along the row the 2D-RLS algorithm is:

Algorithm 4.1.1 (2D-RLS)

(1) Form $\phi^T(m,n)$ using the new data, and form the prediction error

$$\epsilon(m,n) = \begin{cases} y(m,n) - \phi^T(m,n)\hat{\theta}(m-1,n) & \text{for } m > 1 \\ y(m,n) - \phi^T(1,n)\hat{\theta}(W,n-1) & \text{for } m = 1 \end{cases} \quad (4.1.2)$$

(2) Update the new $P(m,n)$ according to Equation 4.1.1.

(3) Form

$$\hat{\theta}(m,n) = \begin{cases} \hat{\theta}(m-1,n) + P(m,n)\phi(m,n)\epsilon(m,n) & \text{for } m > 1 \\ \hat{\theta}(W,n-1) + P(1,n)\phi(1,n)\epsilon(1,n) & \text{for } m = 1 \end{cases}$$

This is directly analogous to the one-dimensional case. But it is clear that jumping from one row to the next requires the assumption of finite boundaries. By restricting the past to a finite width, we are effectively reducing the cost function to

$$J(m,n) = \sum_{i=m-W}^{m-1} \sum_{j} \eta(m-i, n-j)^2$$

Furthermore near $m = 1$ and $m = W$ in some applications the data vector ϕ may be incomplete, since it may include unknown values. This may be overcome by only updating the estimator away from the edges, thus ensuring that the y's are always known in ϕ. It should be noted that if we only estimate away from the edges then the restriction on J becomes more severe—ie we use less information so the convergence of the estimates is likely to be slower.

4.1.2 The extension to ARMA data

As in the one-dimensional case we may extend the estimator to the ARMA model where

$$A(w^{-1}, z^{-1})y(m,n) = C(w^{-1}, z^{-1})e(m,n)$$

with

$$C(w^{-1}, z^{-1}) = \sum_{i=0}^{M_c}\sum_{j=0}^{N_c} c_{i,j} w^{-i} z^{-j} + \sum_{i=1}^{M_c}\sum_{j=1}^{N_c} c_{-i,j} w^{i} z^{-j}$$

The parameter estimator vector is now defined as

$$\hat{\theta}^T(m,n) = (-\hat{a}_{1,0}, ..., -\hat{a}_{M_a,0}, -\hat{a}_{-M_a,1}, ..., -\hat{a}_{M_a,1}, ..., -\hat{a}_{-M_a,N_a}, ... -\hat{a}_{M_a,N_a},$$
$$\hat{c}_{1,0}, ..., \hat{c}_{M_c,0}, \hat{c}_{-M_c,1}, ..., \hat{c}_{M_c,1}, ..., \hat{c}_{-M_c,N_c}, ..\hat{c}_{M_c,N_c})$$

and we must fill the corresponding data vector with estimates of the noise $\hat{e}(m,n)$

$$\phi^T(m,n) = (..., y(m-i, n-j), ..., \hat{e}(m-i, n-j), ...)$$

For 2D-RELS we use the prediction errors $\epsilon(m,n)$ (generated by Equation 4.1.2) as estimates of the noise so that given these modifications the 2D-RELS algorithms is again given by Algorithm 4.1.1. For 2D-AML we use the residuals $\eta(m,n)$ as estimates of the noise. Thus 2D-AML is again given by Algorithm 4.1.1 given the modifications to the parameter estimator and data vectors, and also the calculation of the residual

$$\eta(m,n) = y(m,n) - \phi^T(m,n)\hat{\theta}(m,n)$$

However it should be noted that this extension does introduce difficulties. For 2D-RLS it was observed that in some cases to ensure the data vector always included known values the jump from one row to the next had to occur away from the boundary of known values. For 2D-RELS the total requirement of $\hat{e}$'s may not be known at the edges, although it may be estimated in some other way (for example in the control problem it is reasonable to set these off-edge values to zero). Similarly when starting up the algorithm ϕ must include unknown values of $\hat{e}$ until N_c rows have been scanned (see Fig 4.1.1). Thus the 2D-RELS and 2D-AML algorithms cannot converge to the true values without scanning more than one row.

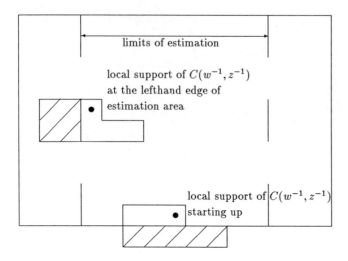

Fig 4.1.1 *Regions where estimates of the noise $e(m,n)$ may be required but are not explicitly calculated in 2D-RELS or 2D-AML.*

4.2 Forgetting strategies

In one dimension the incorporation of a forgetting factor to deal with time-varying parameters is a well-established technique, eg [Astrom et al. 1977]. The constant forgetting factor is a relatively clumsy tool, and more versatile time-varying forgetting factors [Fortescue et al. 1981], [Wellstead and Sanoff 1981] have been developed. Even such sophisticated fault-detection algorithms as those found in [Hagglund 1983] are often based on a forgetting-factor approach.

A form of forgetting was incorporated into the two-dimensional estimator in [Caldas-Pinto 1983] and [Wagner 1987]. However such a forgetting scheme has been criticised, for example in [Angwin 1989], for being essentially one-dimensional. This is because the weighting of previous data is determined by the path of the raster scan rather than the essential two-dimensional structure of the data itself. Here we investigate this failing and present a new forgetting algorithm which may be considered truly two-dimensional, in that it weights data according to their spatial location. The results of simulations in §4.6 (below) and Appendix 4 show that the new algorithm is considerably more versatile

4.2 Forgetting strategies

than those of [Caldas-Pinto 1983] and [Wagner 1987].

Although in [Caldas-Pinto 1983] a time-varying forgetting factor is used and in [Wagner 1987] a directional forgetting factor [Kulhavy and Karny 1984], [Kulhavy 1987] is used, and in both cases the estimator is for ARMA data rather than AR data, the essential philosophy is the same. Namely, to replace Equation 4.1.1 with

$$P(m,n) = \begin{cases} \dfrac{P(m-1,n)}{\lambda}\left[I - \dfrac{\phi(m,n)\phi^T(m,n)P(m-1,n)}{\lambda + \phi^T(m,n)P(m-1,n)\phi(m,n)}\right] & \text{for } m > 1 \\[2ex] \dfrac{P(W,n-1)}{\lambda}\left[I - \dfrac{\phi(1,n)\phi^T(1,n)P(W,n-1)}{\lambda + \phi^T(1,n)P(W,n-1)\phi(1,n)}\right] & \text{for } m = 1 \end{cases}$$

(4.2.1)

where λ is the forgetting factor.

The relation between this forgetting strategy and that for a one-dimensional estimator is obvious. At each step in the *scan* previous information is scaled by a factor λ. Thus effectively we are minimising the cost function

$$J(m,n) = \sum_{i=0}^{m-1}\sum_{j=0}^{\infty} \lambda^{i+jW} \eta(m-i,n-j)^2 + \sum_{i=1}^{W-m}\sum_{j=1}^{\infty} \lambda^{jW-i} \eta(m+i,n-j)^2$$

where W is the width of the estimation scan. Thus the weighting of past data does not correspond to high correlation with the current pixel; for instance any pixel on the current row is given a higher weighting than any pixel on a previous row (see Fig 4.2.1).

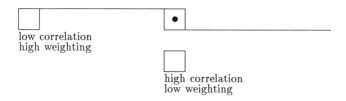

Fig 4.2.1 *Weighting of pixel information in previous forgetting strategies.*

We propose here to introduce a new forgetting strategy which may be considered truly two-dimensional in that data points are weighted according to their proximity to the current pixel. Specifically, when there are no edges to the data, we wish to minimise the cost function

$$J(m,n) = \sum_{i=0}^{\infty}\sum_{j=0}^{\infty} \lambda^i \mu^j \eta(m-i, n-j)^2 + \sum_{i=1}^{\infty}\sum_{j=1}^{\infty} \lambda^i \mu^j \eta(m+i, n-j)^2$$

If there is a finite width to the data then the cost function becomes

$$J(m,n) = \sum_{i=0}^{m-1}\sum_{j=0}^{\infty} \lambda^i \mu^j \eta(m-i, n-j)^2 + \sum_{i=1}^{W-m}\sum_{j=1}^{\infty} \lambda^i \mu^j \eta(m+i, n-j)^2$$

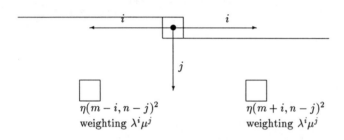

$\eta(m-i, n-j)^2$ weighting $\lambda^i \mu^j$

$\eta(m+i, n-j)^2$ weighting $\lambda^i \mu^j$

Fig 4.2.2 *Weighting of pixel information in the new forgetting strategy.*

Then the derivative of $J(m,n)$ is given by

$$\frac{\partial J(m,n)}{\partial \hat{\theta}} = 2 \sum_{i=0}^{m-1}\sum_{j=0}^{\infty} \lambda^i \mu^j \left[y(m-i, n-j) - \phi^T(m-i, n-j)\hat{\theta}(m,n) \right] \times \phi(m-i, n-j)$$
$$+ 2 \sum_{i=1}^{W-m}\sum_{j=1}^{\infty} \lambda^i \mu^j \left[y(m+i, n-j) - \phi^T(m+i, n-j)\hat{\theta}(m,n) \right] \times \phi(m+i, n-j)$$

So if we introduce

$$Q_l(m,n) = \sum_{i=0}^{m-1}\sum_{j=0}^{\infty} \lambda^i \mu^j \phi(m-i, n-j)\phi^T(m-i, n-j) \qquad (4.2.2)$$

$$Q_r(m,n) = \sum_{i=1}^{W-m}\sum_{j=1}^{\infty} \lambda^i \mu^j \phi(m+i, n-j)\phi^T(m+i, n-j) \qquad (4.2.3)$$

4.2 Forgetting strategies

$$R_l(m,n) = \sum_{i=0}^{m-1} \sum_{j=0}^{\infty} \lambda^i \mu^j \phi(m-i, n-j) y(m-i, n-j) \quad (4.2.4)$$

and

$$R_r(m,n) = \sum_{i=1}^{W-m} \sum_{j=1}^{\infty} \lambda^i \mu^j \phi(m+i, n-j) y(m+i, n-j) \quad (4.2.5)$$

where the supports corresponding to $Q_l(m,n)$ and $Q_r(m,n)$ are shown in Fig 4.2.3, then

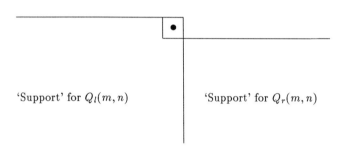

'Support' for $Q_l(m,n)$ 'Support' for $Q_r(m,n)$

Fig 4.2.3 The 'support regions' for $Q_l(m,n)$ and $Q_r(m,n)$.

$$\frac{\partial J}{\partial \hat{\theta}} = -2R_l(m,n) + 2Q_l(m,n)\hat{\theta}(m,n) - 2R_r(m,n) + 2Q_r(m,n)\hat{\theta}(m,n)$$

so

$$\hat{\theta}(m,n) = [Q_l(m,n) + Q_r(m,n)]^{-1} [R_l(m,n) + R_r(m,n)] \quad (4.2.6)$$

We wish to find a recursive formulation for the estimate $\hat{\theta}$. To do so we will find it convenient to consider three cases:

(1) $0 < \lambda < 1$

This is the general case and we will consider it in the next two sections (§4.3 and §4.4). From now on we will refer to this as general two-dimensional forgetting.

(2) $\lambda = 1$

We will refer to this special case as row forgetting and consider it in §4.5.

(3) $\lambda = 0$

Similarly we will refer to this special case as column forgetting and consider it in §4.5 also.

4.3 Two-dimensional forgetting with $0 < \lambda < 1$

4.3.1 An 'Attasi's model' form for two-dimensional forgetting

Here we develop the more general forgetting algorithm where both λ and μ are non-unity. For general λ we have the relationships

$$R_l(m,n) = \sum_{i=0}^{m-1} \sum_{j=0}^{\infty} \lambda^i \mu^j \phi(m-i, n-j) y(m-i, n-j)$$

$$= \lambda R_l(m-1, n) + \sum_{j=0}^{\infty} \mu^j \phi(m, n-j) y(m, n-j) \quad (4.3.1)$$

and

$$R_r(m,n) = \sum_{i=1}^{W-m} \sum_{j=1}^{\infty} \lambda^i \mu^j \phi(m+i, n-j) y(m+i, n-j)$$

$$= \frac{1}{\lambda} R_r(m-1, n) - \sum_{j=1}^{\infty} \mu^j \phi(m+1, n-j) y(m+1, n-j) \quad (4.3.2)$$

Note that the recursive formula for R_l is similar to the one-dimensional case (although a whole column of data is added at each step). However, as we move from left to right the 'support' for R_r decreases in area, and hence the minus sign in the recursive formula. On the other hand the weighting for each component in R_r increases, and hence the multiplication by $\frac{1}{\lambda}$.

Combining Equation 4.3.1 with Equation 4.3.2 gives

$$R_l(m,n) + R_r(m,n) = \lambda R_l(m-1,n) + \frac{1}{\lambda} R_r(m-1,n) + \phi(m,n) y(m,n) \quad (4.3.3)$$

Similarly

$$Q_l(m,n) + Q_r(m,n) = \lambda Q_l(m-1,n) + \frac{1}{\lambda} Q_r(m-1,n) + \phi(m,n) \phi^T(m,n) \quad (4.3.4)$$

It follows that the matrix inversion lemma [Ljung and Soderstrom 1983] cannot be invoked for the update of

$$[Q_l(m,n) + Q_r(m,n)]^{-1}$$

Instead we will derive algorithms for updating Q_l, Q_r, R_l and R_r directly and assume that we have enough computational power to invert $(Q_l + Q_r)$ at each new 'pixel'.

4.3 Two-dimensional forgetting with $0 < \lambda < 1$

We have

$$Q_l(m,n) = \sum_{i=0}^{m-1}\sum_{j=0}^{\infty} \lambda^i \mu^j \phi(m-i, n-j)\phi^T(m-i, n-j)$$

$$= \lambda Q_l(m-1, n) + \sum_{j=0}^{\infty} \mu^j \phi(m, n-j)\phi^T(m, n-j) \quad (4.3.5)$$

and similarly

$$Q_l(m, n-1) = \sum_{i=0}^{m-1}\sum_{j=0}^{\infty} \lambda^i \mu^j \phi(m-i, n-1-j)\phi^T(m-i, n-1-j)$$

$$= \lambda Q_l(m-1, n-1) + \sum_{j=0}^{\infty} \mu^j \phi(m, n-1-j)\phi^T(m, n-1-j) \quad (4.3.6)$$

Combining Equation 4.3.5 with Equation 4.3.6 we obtain

$$Q_l(m,n) - \mu Q_l(m, n-1) = \lambda Q_l(m-1, n) - \lambda\mu Q_l(m-1, n-1)$$
$$+ \phi(m,n)\phi^T(m,n)$$

or

$$Q_l(m,n) = \lambda Q_l(m-1, n) + \mu Q_l(m, n-1) - \lambda\mu Q_l(m-1, n-1)$$
$$+ \phi(m,n)\phi^T(m,n) \quad (4.3.7)$$

or we may write

$$(1 - \lambda w^{-1})(1 - \mu z^{-1})Q_l(m,n) = \phi(m,n)\phi^T(m,n)$$

Similarly

$$Q_r(m,n) = \lambda Q_r(m+1, n) + \mu Q_r(m, n-1) - \lambda\mu Q_r(m+1, n-1)$$
$$+ \lambda\mu \phi(m+1, n-1)\phi^T(m+1, n-1) \quad (4.3.8)$$

or

$$(1 - \lambda w)(1 - \mu z^{-1})Q_r(m,n) = \lambda\mu \phi(m+1, n-1)\phi^T(m+1, n-1)$$

We may also establish

$$R_l(m,n) = \lambda R_l(m-1, n) + \mu R_l(m, n-1) - \lambda\mu R_l(m-1, n-1)$$
$$+ \phi(m,n)y(m,n) \quad (4.3.9)$$

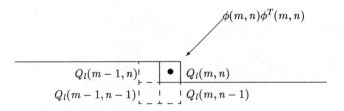

Fig 4.3.1 The update of $Q_l(m,n)$.

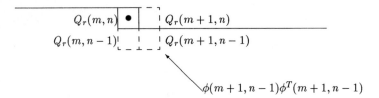

Fig 4.3.2 The update of $Q_r(m,n)$.

and

$$R_r(m,n) = \lambda R_r(m+1,n) + \mu R_r(m,n-1) - \lambda\mu R_r(m+1,n-1)$$
$$+ \lambda\mu\phi(m+1,n-1)y(m+1,n-1) \quad (4.3.10)$$

It is interesting to compare these formulae with Attasi's two-dimensional state-space model [Attasi 1975], [Attasi 1973].

Note that the recursion formulae for Q_r and R_r are non-causal. It is in fact possible to express these formulae in a causal manner, for example

$$Q_r(m,n) = \frac{1}{\lambda}Q_r(m-1,n) - \frac{\mu}{\lambda}Q_r(m-1,n-1) + \mu Q_r(m,n-1)$$
$$- \mu\phi(m,n-1)\phi^T(m,n-1)$$

However, if we return to our original definition (Equation 4.2.3) it becomes clear that we cannot build up Q_r from left to right in this manner. Effectively we would have to

4.3 Two-dimensional forgetting with $0 < \lambda < 1$

start up with complete knowledge of Q_r as our initial condition, and lose information at each step. Instead, by recursing from right to left, each step adds to our knowledge of Q_r. Furthermore, since the new information required for $Q_r(m,n)$ is data from the $(n-1)$'th row, we may perform all the 'non-causal' recursions for each row in one fell swoop at the end of the previous row, without requiring knowledge of unknown future data. Our algorithm is then:

Algorithm 4.3.1 (Basic two-dimensional forgetting factors)

(1) Scan along the row and perform the following operations

    (i) Form $\phi(m,n)$ using the available data, and using the residuals $\eta(m-i, n-j)$ as estimates for the noise $e(m-i, n-j)$.

    (ii) Update to the new $Q_l(m,n)$ and $R_l(m,n)$

    (iii) Form

$$\hat{\theta}(m,n) = [Q_l(m,n) + Q_r(m,n)]^{-1} [R_l(m,n) + R_r(m,n)] \quad (4.3.11)$$

    (Note that the Q_r's here have already been formed at the end of the previous row).

    (iv) Calculate the current residual from

$$\eta(m,n) = y(m,n) - \phi^T(m,n)\hat{\theta}(m,n)$$

(2) At the end of the row update all the $Q_r(W-i, n+1)$'s and $R_r(W-i, n+1)$'s for the next row.

(3) Repeat steps (i), (ii), (iii) and (iv) for the row $n+1$, etc.

Note that this form is equivalent to AML (as opposed to RELS) in that we use the residuals as estimates for the noise rather than the prediction errors. In two dimensions this makes good sense since if a parameter change occurs in the vertical direction then $\hat{\theta}(m-1, n)$ is not necessarily a close approximation to $\hat{\theta}(m,n)$. Using AML avoids the necessity of using $\hat{\theta}(m-1, n)$ explicitly in the calculation of $\hat{\theta}(m,n)$.

We have already made the assumption that we have enough computational power to invert the information matrix $(Q_l + Q_r)$ at each step (the dimension of the information matrix is the number of parameters in θ). Although we have to update a whole row of Q_r's and R_r's at once, this involves only matrix addition and multiplication by scalars,

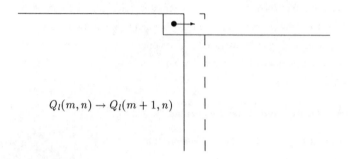

Fig 4.3.3 *The updating of $Q_l(m,n)$ with the scan.*

so should not add significantly to the computational burden. However, the memory requirement of the algorithm is large, since at any one time the values of Q_l, Q_r, R_l, R_r and ϕ must be stored for two previous rows.

Because we explicitly invert $(Q_l + Q_r)$ at each step, any algorithm implementing the two-dimensional forgetting factor must include some logic to prevent the inversion when $(Q_l + Q_r)$ is singular. In particular, this will be the case when starting up. In the simulations of §4.6 (below) pseudo-random data were substituted for $\hat{e}(m,n)$ at each step until the information matrix $(Q_l + Q_r)$ became non-singular.

However this algorithm is unsuitable on two counts. Firstly the subtraction of data in Equations 4.3.7, 4.3.8, 4.3.9 and 4.3.10 is likely to lead to numerical inaccuracy. Secondly the matrix inversion (Equation 4.3.11) is crude and may be considerably improved upon.

4.3.2 A 'Roesser's model' form for two-dimensional forgetting

In §4.3.1 we remarked that the formulae for the update of Q_l, Q_r, R_l and R_r (Equations 4.3.7, 4.3.8, 4.3.9 and 4.3.10) could be compared with the state-space model of Attasi [Attasi 1975], [Attasi 1973]. It is well known [Kaczorek 1985] that this model is a special case of the more general Roesser state-space model [Roesser 1975]. However the states in the Attasi model are not those of the Roesser model. It turns out that we can reformulate the algorithm in a manner analogous to Roesser's model and obtain a form which does not require subtraction.

First consider the state-space models. Attasi's model is of the form [Attasi 1975],

4.3 Two-dimensional forgetting with $0 < \lambda < 1$

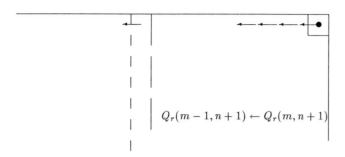

Fig 4.3.4 *The updating of a row of $Q_r(m,n)$'s.*

[Attasi 1973]

$$\bar{x}(i+1, j+1) = A_1^a \bar{x}(i+1, j) + A_2^a \bar{x}(i, j+1) - A_1^a A_2^a \bar{x}(i, j) + B^a u(i, j)$$

$$y(i, j) = C^a \bar{x}(i, j)$$

where

$$A_1^a A_2^a = A_2^a A_1^a$$

Meanwhile Roesser's model [Roesser 1975] is of the form

$$\begin{bmatrix} x^h(i+1, j) \\ x^v(i, j+1) \end{bmatrix} = \begin{bmatrix} A_{11}^r & A_{12}^r \\ A_{21}^r & A_{22}^r \end{bmatrix} \begin{bmatrix} x^h(i, j) \\ x^v(i, j) \end{bmatrix} + \begin{bmatrix} B_1^r \\ B_2^r \end{bmatrix} u(i, j)$$

$$y(i, j) = \begin{bmatrix} C_1^r & C_2^r \end{bmatrix} \begin{bmatrix} x^h(i, j) \\ x^v(i, j) \end{bmatrix}$$

Here $x^h(i, j)$ and $x^v(i, j)$ are respectively the horizontal and vertical states.

The connection between the two models is as follows [Kaczorek 1985]. Given Attasi's model choose the Roesser states as

$$x^h(i, j) = \bar{x}(i, j+1) - A_1^a \bar{x}(i, j)$$

and

$$x^v(i, j) = \bar{x}(i, j)$$

Then we have the relationships

$$x^h(i+1,j) = \bar{x}(i+1,j+1) - A_1^a \bar{x}(i+1,j)$$

$$= A_2^a \bar{x}(i,j+1) - A_1^a A_2^a \bar{x}(i,j) + B^a u(i,j)$$

$$= A_2^a x^h(i,j) + B^a u(i,j)$$

and

$$x^v(i,j+1) = \bar{x}(i,j+1)$$

$$= x^h(i,j) + A_1^a \bar{x}(i,j)$$

$$= x^h(i,j) + A_1^a x^v(i,j)$$

Hence we have the Roesser model

$$\begin{bmatrix} x^h(i+1,j) \\ x^v(i,j+1) \end{bmatrix} = \begin{bmatrix} A_2^a & 0 \\ 1 & A_1^a \end{bmatrix} \begin{bmatrix} x^h(i,j) \\ x^v(i,j) \end{bmatrix} + \begin{bmatrix} B^a \\ 0 \end{bmatrix} u(i,j)$$

$$y(i,j) = \begin{bmatrix} 0 & C \end{bmatrix} \begin{bmatrix} x^h(i,j) \\ x^v(i,j) \end{bmatrix}$$

In an analogous manner we can define new matrices Q_l^h and Q_l^v in terms of Q_l. Specifically if we define

$$Q_l^h(m,n) = Q_l(m,n) - \mu Q_l(m,n-1)$$

$$= \sum_{i=0}^{m-1} \lambda^i \phi(m-i,n) \phi^T(m-i,n)$$

and

$$Q_l^v(m,n) = Q_l(m,n)$$

then we obtain the recursion laws

$$Q_l^h(m,n) = \lambda Q_l(m-1,n) + \phi(m,n) \phi^T(m,n)$$

4.3 Two-dimensional forgetting with $0 < \lambda < 1$

and

$$Q_l^v(m,n) = \mu Q_l^v(m, n-1) + Q_l^h(m,n)$$

with the boundary value

$$Q_l^h(1,n) = \phi(1,n)\phi^T(1,n)$$

Similarly for the righthand side we can define

$$\begin{aligned} Q_r^h(m,n) &= Q_r(m,n) - \mu Q_r(m, n-1) \\ &= \sum_{i=1}^{W-m} \lambda^i \phi(m+i, n-1)\phi^T(m+i, n-1) \end{aligned}$$

and

$$Q_r^v(m,n) = Q_r(m,n)$$

We then obtain the recursion laws

$$Q_r^h(m,n) = \lambda Q_r(m+1,n) + \lambda\mu \phi(m+1, n-1)\phi^T(m+1, n-1)$$

and

$$Q_r^v(m,n) = \mu Q_r^v(m, n-1) + Q_r^h(m,n)$$

with the boundary value

$$Q_r^h(W,n) = 0$$

Furthermore we can combine Q_l^v and Q_r^v to form

$$\begin{aligned} Q^v(m,n) &= Q_l^v(m,n) + Q_r^v(m,n) \\ &= \sum_{i=0}^{m-1} \sum_{j=0}^{\infty} \lambda^i \mu^j \phi(m-i, n-j)\phi^T(m-i, n-j) \\ &\quad + \sum_{i=1}^{W-m} \sum_{j=1}^{\infty} \lambda^i \mu^j \phi(m+i, n-j)\phi^T(m+i, n-j) \end{aligned}$$

with the recursion law

$$Q^v(m,n) = \mu Q^v(m, n-1) + Q_l^h(m,n) + Q_r^h(m,n)$$

If we define $R_l^h(m,n)$, $R_r^h(m,n)$ and $R^v(m, n-1)$ similarly then we can obtain the following algorithm for estimating $\hat{\theta}(m,n)$ which does not involve subtraction and its associated numerical difficulties:

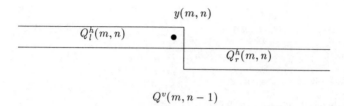

Fig 4.3.5 $Q_l^h(m,n)$, $Q_r^h(m,n)$ and $Q^v(m,n-1)$.

Algorithm 4.3.2 (Two-dimensional forgetting without subtraction: first form)

(1) Scan along the row and perform the following operations

   (i) Form $\phi(m,n)$ using the available data.

   (ii) Form
$$Q_l^h(m,n) = \lambda Q_l^h(m-1,n) + \phi(m,n)\phi^T(m,n)$$
and
$$R_l^h(m,n) = \lambda R_l^h(m-1,n) + \phi(m,n)y(m,n)$$

   (iii) Form
$$Q^v(m,n) = \mu Q^v(m,n-1) + Q_l^h(m,n) + Q_r^h(m,n)$$
and
$$R^v(m,n) = \mu R^v(m,n-1) + R_l^h(m,n) + R_r^h(m,n)$$

   (iv) Form
$$\hat{\theta}(m,n) = [Q^v(m,n)]^{-1} R^v(m,n) \qquad (4.3.12)$$

   (v) Form
$$\eta(m,n) = y(m,n) - \phi^T(m,n)\hat{\theta}(m,n)$$

(2) At the end of each row increase n by 1 and form the following for the whole row:
$$Q_r^h(m,n) = \lambda Q_r^h(m+1,n) + \lambda\mu\phi(m+1,n-1)\phi^T(m+1,n-1)$$

4.3 Two-dimensional forgetting with $0 < \lambda < 1$

and

$$R_r^h(m,n) = \lambda R_r^h(m+1,n) + \lambda\mu\phi(m+1, n-1)y(m+1, n-1)$$

Note that on the return scan we have only a series of rank-one updates. On the other hand we must store $Q^v(m,n)$, $R^v(m,n)$ $Q_r^h(m,n)$ and $R_r^h(m,n)$ for a whole row at a time. We can modify the algorithm so that only $Q^v(m,n)$ and $R^v(m,n)$ need be stored, at the cost of more computation on the return scan. Specifically:

Algorithm 4.3.3 (Two-dimensional forgetting without subtraction: second form)

(1) Scan along the row and perform the following operations:

  (i) Form $\phi(m,n)$ using the available data.

  (ii) Form

$$Q_l^h(m,n) = \lambda Q_l^h(m-1,n) + \phi(m,n)\phi^T(m,n)$$

  and

$$R_l^h(m,n) = \lambda R_l^h(m-1,n) + \phi(m,n)y(m,n)$$

  (iii) Form

$$Q^v(m,n) = Q'^v(m,n) + Q_l^h(m,n)$$

  and

$$R^v(m,n) = R'^v(m,n) + R_l^h(m,n)$$

  (iv) Form

$$\hat{\theta}(m,n) = [Q^v(m,n)]^{-1} R^v(m,n) \qquad (4.3.13)$$

  (v) Form

$$\eta(m,n) = y(m,n) - \phi^T(m,n)\hat{\theta}(m,n)$$

(2) At the end of each row increase n by 1 and form the following for the whole row:

  (i)

$$Q_r^h(m,n) = \lambda Q_r^h(m+1,n) + \lambda\mu\phi(m+1, n-1)\phi^T(m+1, n-1)$$

and
$$R_r^h(m,n) = \lambda R_r^h(m+1,n) + \lambda\mu\phi(m+1,n-1)y(m+1,n-1)$$

(ii) $Q'^v(m,n) = \mu Q'^v(m,n-1) + Q_r^h(m,n)$
and
$$R'^v(m,n) = \mu R'^v(m,n-1) + R_r^h(m,n)$$

4.4 Two-dimensional forgetting in factored form

When implementing recursive least squares and its related estimation techniques it is well known [Lawson and Hanson 1974] that algorithms based on the matrix inversion lemma can be numerically ill-conditioned. It is common practice to implement them in a form where the covariance matrix is in factored form and where square-root calculations are avoided. In particular the Bierman U/D factorisation method [Bierman 1977] is popular. The parameter estimation algorithms for two-dimensional processes we have used which are based on the matrix inversion lemma can similarly be implemented using the Bierman U/D factorisation. However this form of recursion is not used for the two-dimensional forgetting factors. Instead the information matrix $Q^v(m,n)$ is required to be inverted at each step (Equations 4.3.12 and 4.3.13); as it stands this is clearly an unsatisfactory solution. Here we suggest the use of QR factorisation to improve the computational speed and numerical stability of the algorithms.

In this section for clarity we will adopt the notational convention that a quantity doubly underlined $\underline{\underline{X}}$ is a matrix while a quantity singly underlined $\underline{X}$ is a vector. A quantity with no underlining is in general a scalar.

Suppose we have the familiar least squares solution

$$\underline{\hat{\theta}} = \left[\underline{\underline{X}}^T \underline{\underline{X}}\right]^{-1} \underline{\underline{X}}^T \underline{Y}$$

Then if we can decompose

$$\underline{\underline{X}} = \underline{\underline{Q}}\,\underline{\underline{R}} \qquad (4.4.1)$$

where $\underline{\underline{Q}}$ is orthogonal and $\underline{\underline{R}}$ is square upper triangular then

$$\underline{\hat{\theta}} = \left[\underline{\underline{R}}^T \underline{\underline{Q}}^T \underline{\underline{Q}}\, \underline{\underline{R}}\right]^{-1} \underline{\underline{R}}^T \underline{\underline{Q}}^T \underline{Y}$$

4.4 Two-dimensional forgetting in factored form

$$= \left[\underline{\underline{R}}^T \underline{\underline{R}}\right]^{-1} \underline{\underline{R}}^T \underline{\underline{Q}}^T \underline{Y}$$

$$= \underline{\underline{R}}^{-1} \underline{\underline{Q}}^T \underline{Y} \qquad (4.4.2)$$

Since $\underline{\underline{R}}$ is upper triangular $\hat{\underline{\theta}}$ may now be found by simple back substitution.

Let us define $\mathcal{R}$ to be the transformation such that for any matrix $\underline{\underline{X}}$ with such a decomposition as in Equation 4.4.1

$$\mathcal{R}\left[\underline{\underline{X}}\right] = \underline{\underline{Q}}^T \underline{\underline{X}}$$

$$= \underline{\underline{R}} \qquad (4.4.3)$$

where $\underline{\underline{Q}}$ is orthogonal and $\underline{\underline{R}}$ is upper right triangular. Such transforms can be used recursively in one dimension by using the following standard result:

Theorem 4.4.1 *Suppose*

$$\hat{\underline{\theta}} = \left[\underline{\underline{X}}_1^T \underline{\underline{X}}_1 + \underline{\phi}\,\underline{\phi}^T\right]^{-1} \left[\underline{\underline{X}}_1^T \underline{Y}_1 + \underline{\phi} y\right]$$

and we already know

$$\begin{bmatrix} \underline{\underline{R}}_1 & \underline{S}_1 \\ 0 & v_1 \end{bmatrix} = \mathcal{R}\begin{bmatrix} \underline{\underline{X}}_1 & \underline{Y}_1 \end{bmatrix}$$

Then

$$\hat{\underline{\theta}} = \underline{\underline{R}}^{-1} \underline{S}$$

where

$$\begin{bmatrix} \underline{\underline{R}} & \underline{S} \\ 0 & v \end{bmatrix} = \mathcal{R} \begin{bmatrix} \underline{\underline{R}}_1 & \underline{S}_1 \\ 0 & v_1 \\ \underline{\phi}^T & y \end{bmatrix}$$

Proof *See for example [Golub and van Loan 1989].*

For our recursion over a two-dimensional data field we will need the more general result:

Theorem 4.4.2 *Suppose*

$$\hat{\underline{\theta}} = \left[\lambda_1 \underline{\underline{X}}_1^T \underline{\underline{X}}_1 + \lambda_2 \underline{\underline{X}}_2^T \underline{\underline{X}}_2\right]^{-1} \left[\lambda_1 \underline{\underline{X}}_1^T \underline{Y}_1 + \lambda_2 \underline{\underline{X}}_2^T \underline{Y}_2\right]$$

where

$$\lambda_1, \lambda_2 \geq 0$$

and we already know

$$\begin{bmatrix} \underline{\underline{R}}_i & \underline{S}_i \\ 0 & v_i \end{bmatrix} = \mathcal{R} \begin{bmatrix} \underline{\underline{X}}_i & \underline{Y}_i \end{bmatrix} \text{ for } i = 1, 2$$

Then we have the two relations

$$\left[\lambda_1 \underline{\underline{X}}_1^T \underline{\underline{X}}_1 + \lambda_2 \underline{\underline{X}}_2^T \underline{\underline{X}}_2\right] = \underline{\underline{R}}^T \underline{\underline{R}}$$

and

$$\left[\lambda_1 \underline{\underline{X}}_1^T \underline{Y}_1 + \lambda_2 \underline{\underline{X}}_2^T \underline{Y}_2\right] = \underline{\underline{R}}^T \underline{S}$$

and hence

$$\hat{\underline{\theta}} = \underline{\underline{R}}^{-1} \underline{S}$$

where

$$\begin{bmatrix} \underline{\underline{R}} & \underline{S} \\ 0 & v \end{bmatrix} = \mathcal{R} \begin{bmatrix} \lambda_1 \underline{\underline{R}}_1 & \lambda_1 \underline{S}_1 \\ 0 & \lambda_1 v_1 \\ \lambda_2 \underline{\underline{R}}_2 & \lambda_2 \underline{S}_2 \\ 0 & \lambda_2 v_2 \end{bmatrix}$$

Proof *See Appendix 1.*

Unlike the one-dimensional case each update has a rank greater than one. Hence the obvious choice for such a transformation is to use successive Householder transformations [Wilkinson 1977]. Given a vector $\underline{x}$ a single Householder transformation is performed by pre-multiplying by the orthogonal matrix

$$\underline{\underline{P}}_x = \underline{\underline{I}} - \frac{2\underline{v}\,\underline{v}^T}{\underline{v}^T \underline{v}}$$

4.4 Two-dimensional forgetting in factored form

where

$$\underline{v} = \underline{x} \pm \left(\underline{x}^T \underline{x}\right)^{\frac{1}{2}} \underline{e}_1$$

and $\underline{e}_1$ is a unit vector whose first entry is 1 and all others 0. It is then easy to see that

$$\underline{\underline{P}}_{\underline{x}} \underline{x} = \mp \left(\underline{x}^T \underline{x}\right)^{\frac{1}{2}} \underline{e}_1$$

Thus given an $(m \times n)$ matrix we can reduce it to upper right triangular form using m successive Householder transformations as follows. Suppose we denote the non-zero elements of a matrix with ×'s and begin with the matrix

$$\underline{\underline{X}} = \left[\underline{x}_1^0 \underline{x}_2^0 \ldots \underline{x}_m^0\right]$$

where each $\underline{x}_i^0$ is an n-vector. Then

$$\underline{\underline{P}}_{\underline{x}_1^0} \underline{\underline{X}} = \begin{bmatrix} \times & \times & \ldots & \times \\ 0 & \underline{x}_2^1 & \ldots & \underline{x}_m^1 \end{bmatrix}$$

where each $\underline{x}_i^1$ is an $(n-1)$-vector. Our second transformation is then

$$\begin{bmatrix} 1 & 0 \\ 0 & \underline{\underline{P}}_{\underline{x}_2^1} \end{bmatrix} \underline{\underline{P}}_{\underline{x}_1^0} \underline{\underline{X}} = \begin{bmatrix} \times & \times & \times & \ldots & \times \\ 0 & \times & \times & \ldots & \times \\ 0 & 0 & \underline{x}_3^2 & \ldots & \underline{x}_m^2 \end{bmatrix}$$

and our third

$$\begin{bmatrix} 1 & 0 & 0 \\ 0 & 1 & 0 \\ 0 & 0 & \underline{\underline{P}}_{\underline{x}_3^2} \end{bmatrix} \begin{bmatrix} 1 & 0 \\ 0 & \underline{\underline{P}}_{\underline{x}_2^1} \end{bmatrix} \underline{\underline{P}}_{\underline{x}_1^0} \underline{\underline{X}} = \begin{bmatrix} \times & \times & \times & \times & \ldots & \times \\ 0 & \times & \times & \times & \ldots & \times \\ 0 & 0 & \times & \times & \ldots & \times \\ 0 & 0 & 0 & \underline{x}_4^3 & \ldots & \underline{x}_m^3 \end{bmatrix}$$

(4.4.4)

We continue in this manner until we have an upper right triangular matrix. See [Wilkinson 1977] or [Golub and van Loan 1989] for more details. We denote the such a succession of Householder transformations as $\mathcal{H}$ so that

$$\mathcal{H}\left[\underline{\underline{X}}\right] = \underline{\underline{R}}$$

where $\underline{\underline{R}}$ is upper right triangular and there exists some orthogonal $\underline{\underline{Q}}$ such that

$$\underline{\underline{X}} = \underline{\underline{Q}} \, \underline{\underline{R}}$$

Then we can restate the two-dimensional forgetting-factors algorithm using Householder transformations.

Algorithm 4.4.1 (Factored form of Algorithm 4.3.2 using Householder transformations)

(1) Scan along the row and perform the following operations.

   (i) Form $\phi(m,n)$ using the available data.

   (ii) Form

   $$\begin{bmatrix} \underline{\underline{R}}_l^h(m,n) & \underline{S}_l^h(m,n) \\ 0 & v_l^h(m,n) \end{bmatrix} = \mathcal{H} \begin{bmatrix} \lambda^{\frac{1}{2}}\underline{\underline{R}}_l^h(m-1,n) & \lambda^{\frac{1}{2}}\underline{S}_l^h(m-1,n) \\ 0 & \lambda^{\frac{1}{2}}v_l^h(m-1,n) \\ \phi^T(m,n) & y(m,n) \end{bmatrix} \quad (4.4.5)$$

   (iii) Form

   $$\begin{bmatrix} \underline{\underline{R}}^v(m,n) & \underline{S}^v(m,n) \\ 0 & v^v(m,n) \end{bmatrix} = \mathcal{H} \begin{bmatrix} \mu^{\frac{1}{2}}\underline{\underline{R}}^v(m,n-1) & \mu^{\frac{1}{2}}\underline{S}^v(m,n-1) \\ 0 & \mu^{\frac{1}{2}}v^v(m,n-1) \\ \underline{\underline{R}}_l^h(m,n) & \underline{S}_l^h(m,n) \\ 0 & v_l^h(m,n) \\ \underline{\underline{R}}_r^h(m,n) & \underline{S}_r^h(m,n) \\ 0 & v_r^h(m,n) \end{bmatrix} \quad (4.4.6)$$

   (iv) Form

   $$\hat{\underline{\theta}}(m,n) = \left[\underline{\underline{R}}^v(m,n)\right]^{-1} \underline{S}^v(m,n)$$

   (v) Form

   $$\eta(m,n) = y(m,n) - \phi(m,n)\hat{\underline{\theta}}(m,n)$$

(2) At the end of each row increase n by 1 and form the following for the whole row:

$$\begin{bmatrix} \underline{\underline{R}}_r^h(m,n) & \underline{S}_r^h(m,n) \\ 0 & v_r^h(m,n) \end{bmatrix} = \mathcal{H} \begin{bmatrix} \lambda^{\frac{1}{2}}\underline{\underline{R}}_r^h(m+1,n) & \lambda^{\frac{1}{2}}\underline{S}_r^h(m+1,n) \\ 0 & \lambda^{\frac{1}{2}}v_r^h(m+1,n) \\ \lambda^{\frac{1}{2}}\mu^{\frac{1}{2}}\phi^T(m+1,n-1) & \lambda^{\frac{1}{2}}\mu^{\frac{1}{2}}y(m+1,n-1) \end{bmatrix} \quad (4.4.7)$$

4.4 Two-dimensional forgetting in factored form

The alternative to using Householder transformations is to use Givens rotations. A Givens matrix is defined [Golub and van Loan 1989] to be of the form

$$\underline{\underline{G}} = \begin{bmatrix} 1 & \cdots & 0 & \cdots & 0 & \cdots & 0 \\ \vdots & \ddots & \vdots & & \vdots & & \vdots \\ 0 & \cdots & c & \cdots & s & \cdots & 0 \\ \vdots & & \vdots & \ddots & \vdots & & \vdots \\ 0 & \cdots & -s & \cdots & c & \cdots & 0 \\ \vdots & & \vdots & & \vdots & \ddots & \vdots \\ 0 & \cdots & 0 & \cdots & 0 & \cdots & 1 \end{bmatrix} \begin{matrix} \\ \\ i \\ \\ k \\ \\ \\ \end{matrix}$$

$$ i k$$

Then if

$$\underline{y} = \underline{\underline{G}}^T \underline{x}$$

and

$$c = \frac{x(i)}{(x(i)^2 + x(k)^2)^{\frac{1}{2}}} \text{ and } s = \frac{-x(k)}{(x(i)^2 + x(k)^2)^{\frac{1}{2}}}$$

then

$$y(j) = \begin{cases} cx(i) - sx(k) & \text{for } j = i \\ 0 & \text{for } j = k \\ y(j) & \text{otherwise} \end{cases}$$

Thus by a series of such rotations we can reduce a matrix to upper right triangular form. Since a single rotation only sets one element of a matrix to zero (as opposed to a whole column using Householder transformations) many more operations are required to reduce a matrix to upper right triangular form by Givens rotations than by Householder transformations. However the advantage of using Givens rotations is that they can be computed in 'fast' form—that is without calculating square roots [Gentleman 1973]. Briefly, given $\underline{\underline{X}}$ and a diagonal matrix $\underline{\underline{D}}_0$ (which may of course be the identity matrix) a QR decomposition using fast Givens rotations denoted $\mathcal{G}$ computes

$$\mathcal{G}\left\{\underline{\underline{D}}_0, \underline{\underline{X}}\right\} = \left\{\underline{\underline{D}}, \underline{\underline{R}}\right\}$$

where $\underline{\underline{D}}$ is diagonal, $\underline{\underline{R}}$ is upper right triangular with unit (or zero) diagonal elements and there is some orthogonal $\underline{\underline{Q}}$ such that

$$\underline{\underline{D}}_0^{\frac{1}{2}} \underline{\underline{X}} = \underline{\underline{Q}} \, \underline{\underline{D}}^{\frac{1}{2}} \underline{\underline{R}}$$

Fast Givens transformations have the further advantage that forgetting factors are more easily incorporated [Gentleman 1973]. For a general $(n \times n)$-diagonal matrix $\underline{\underline{D}}$ denote the $(n-1 \times n-1)$-diagonal matrix formed from its first $(n-1)$ rows and columns as $\underline{\underline{\bar{D}}}$. Then Theorem 4.4.2 becomes:

Theorem 4.4.3 *Suppose*

$$\hat{\theta} = \left[\lambda_1 \underline{X}_1^T \underline{X}_1 + \lambda_2 \underline{X}_2^T \underline{X}_2\right]^{-1} \left[\lambda_1 \underline{X}_1^T \underline{Y}_1 + \lambda_2 \underline{X}_2^T \underline{Y}_2\right]$$

and we already know

$$\left\{\underline{\underline{D}}_i, \begin{bmatrix} \underline{R}_i & \underline{S}_i \\ 0 & v_i \end{bmatrix}\right\} = \mathcal{G}\left\{\underline{\underline{I}}, \begin{bmatrix} \underline{X}_i & \underline{Y}_i \end{bmatrix}\right\} \text{ for } i = 1, 2$$

Then we have the two relations

$$\left[\lambda_1 \underline{X}_1^T \underline{X}_1 + \lambda_2 \underline{X}_2^T \underline{X}_2\right] = \underline{R}^T \underline{\bar{D}}\, \underline{R}$$

and

$$\left[\lambda_1 \underline{X}_1^T \underline{Y}_1 + \lambda_2 \underline{X}_2^T \underline{Y}_2\right] = \underline{R}^T \underline{\bar{D}}\, \underline{S}$$

and hence

$$\hat{\theta} = \underline{R}^{-1} \underline{S}$$

where

$$\left\{\underline{\underline{D}}, \begin{bmatrix} \underline{R} & \underline{S} \\ 0 & v \end{bmatrix}\right\} = \mathcal{G}\left\{\begin{bmatrix} \lambda_1 \underline{\underline{D}}_1 & 0 \\ 0 & \lambda_2 \underline{\underline{D}}_2 \end{bmatrix}, \begin{bmatrix} \underline{R}_1 & \underline{S}_1 \\ 0 & v_1 \\ \underline{R}_2 & \underline{S}_2 \\ 0 & v_2 \end{bmatrix}\right\}$$

Proof *See Appendix 1.*

Hence we can implement the algorithm for two-dimensional forgetting factors using fast Givens rotations:

Algorithm 4.4.2 (Factored form of Algorithm 4.3.2 using fast Givens rotations)

(1) Scan along the row and perform the following operations.

4.4 Two-dimensional forgetting in factored form

(i) Form $\underline{\phi}(m,n)$ using the available data.

(ii) Form

$$\left\{\underline{\underline{D}}_l^h(m,n), \begin{bmatrix} \underline{\underline{R}}_l^h(m,n) & \underline{S}_l^h(m,n) \\ 0 & v_l^h(m,n) \end{bmatrix}\right\} =$$

$$\mathcal{G}\left\{\begin{bmatrix} \lambda \underline{\underline{D}}_l^h(m-1,n) & 0 \\ 0 & 1 \end{bmatrix}, \begin{bmatrix} \underline{\underline{R}}_l^h(m-1,n) & \underline{S}_l^h(m-1,n) \\ 0 & v_l^h(m-1,n) \\ \underline{\phi}^T(m,n) & y(m,n) \end{bmatrix}\right\}$$

(4.4.8)

(iii) Form

$$\left\{\underline{\underline{D}}^v(m,n), \begin{bmatrix} \underline{\underline{R}}^v(m,n) & \underline{S}^v(m,n) \\ 0 & v^v(m,n) \end{bmatrix}\right\} =$$

$$\mathcal{G}\left\{\underline{\underline{D}}'(m,n), \begin{bmatrix} \underline{\underline{R}}^v(m,n-1) & \underline{S}^v(m,n-1) \\ 0 & v^v(m,n-1) \\ \underline{\underline{R}}_l^h(m,n) & \underline{S}_l^h(m,n) \\ 0 & v_l^h(m,n) \\ \underline{\underline{R}}_r^h(m,n) & \underline{S}_r^h(m,n) \\ 0 & v_r^h(m,n) \end{bmatrix}\right\}$$

(4.4.9)

where

$$\underline{\underline{D}}'(m,n) = \begin{bmatrix} \mu \underline{\underline{D}}^v(m,n-1) & 0 & 0 \\ 0 & \underline{\underline{D}}_l^h(m,n) & 0 \\ 0 & 0 & \underline{\underline{D}}_r^h(m,n) \end{bmatrix}$$

(iv) Form

$$\hat{\underline{\theta}}(m,n) = \left[\underline{\underline{R}}^v(m,n)\right]^{-1} \underline{S}^v(m,n)$$

(v) Form

$$\eta(m,n) = y(m,n) - \underline{\phi}^T(m,n)\hat{\underline{\theta}}(m,n)$$

(2) At the end of each row increase n by 1 and form the following for the whole row:

$$\left\{\underline{\underline{D}}_r^h(m,n), \begin{bmatrix} \underline{\underline{R}}_r^h(m,n) & \underline{S}_r^h(m,n) \\ 0 & v_r^h(m,n) \end{bmatrix}\right\} =$$

$$\mathcal{G}\left\{\begin{bmatrix} \lambda \underline{\underline{D}}_r^h(m+1,n) & 0 \\ 0 & \lambda\mu \end{bmatrix}, \begin{bmatrix} \underline{\underline{R}}_r^h(m+1,n) & \underline{S}_r^h(m+1,n) \\ 0 & v_r^h(m+1,n) \\ \underline{\phi}^T(m+1,n-1) & y(m+1,n-1) \end{bmatrix}\right\}$$

(4.4.10)

4.5 Row and column forgetting

In §4.2 we derived a general expression (Equations 4.2.2 to 4.2.5 and Equation 4.2.6) for weighted least squares parameter estimation. §4.3 and §4.4 were devoted to finding suitable algorithmic forms for its implementation in the general case. However we noted in §4.2 that there were two particular cases (when $\lambda = 1$ and $\lambda = 0$ respectively) which deserved separate attention. We will term these row forgetting and column forgetting respectively.

4.5.1 Row forgetting

Recall (Equation 4.2.6) that the estimator $\hat{\theta}(m,n)$ is given by

$$\hat{\theta}(m,n) = [Q_l(m,n) + Q_r(m,n)]^{-1} [R_l(m,n) + R_r(m,n)]$$

where we have the relations (Equations 4.3.3 and 4.3.4)

$$R_l(m,n) + R_r(m,n) = \lambda R_l(m-1,n) + \frac{1}{\lambda} R_r(m-1,n) + \phi(m,n)y(m,n)$$

(4.5.1)

and

$$Q_l(m,n) + Q_r(m,n) = \lambda Q_l(m-1,n) + \frac{1}{\lambda} Q_r(m-1,n) + \phi(m,n)\phi^T(m,n)$$

(4.5.2)

We have seen that in general we cannot invoke the matrix inversion lemma for the update of the covariance matrix. However when $\lambda = 1$ this reduces to the case in §4.1.1 when scanning along each row. Hence we can use Algorithm 4.1.1 for updating along the row. When jumping from one row to the next, assume again we begin and end each row at $m = 1$ and $m = W$ respectively. Note that we have the boundary conditions

$$R_l(1, n+1) = \sum_{j=0}^{\infty} \mu^j \phi(1, n-j) y(1, n-j)$$

and

$$R_r(W, n) = 0$$

4.5 Row and column forgetting

Thus

$$R_l(1, n+1) + R_r(1, n+1) = \sum_{j=0}^{\infty} \mu^j \phi(1, n-j) y(1, n-j)$$
$$+ \sum_{i=1}^{W-1} \sum_{j=1}^{\infty} \mu^j \phi(1+i, n+1-j) y(1+i, n+1-j)$$
$$= \phi(1, n+1) y(1, n+1) + \sum_{i=0}^{W-1} \sum_{j=1}^{\infty} \mu^j \phi(1+i, n+1-j) y(1+i, n+1-j)$$
$$= \phi(1, n+1) y(1, n+1) + \mu R_l(W, n)$$
$$= \phi(1, n+1) y(1, n+1) + \mu [R_l(W, n) + R_r(W, n)] \quad (4.5.3)$$

Similarly we have

$$Q_l(1, n+1) + Q_r(1, n+1) = \phi(1, n+1) \phi^T(1, n+1)$$
$$+ \mu [Q_l(W, n) + Q_r(W, n)] \quad (4.5.4)$$

So again we may invoke Algorithm 4.1.1, except we replace Equation 4.1.1 by

$$P(m, n) = \begin{cases} P(m-1, n) \left[I - \dfrac{\phi(m,n)\phi^T(m,n)P(m-1,n)}{1 + \phi^T(m,n)P(m-1,n)\phi(m,n)} \right] & \text{for } m > 1 \\[2ex] \dfrac{P(W, n-1)}{\mu} \left[I - \dfrac{\phi(1,n)\phi^T(1,n)P(W,n-1)}{\mu + \phi^T(1,n)P(W,n-1)\phi(1,n)} \right] & \text{for } m = 1 \end{cases}$$
(4.5.5)

Note that this case with $\lambda = 1$ corresponds to the case where we only update when we jump from row to row. In other words each 'pixel' on a given row is given an equal weighting, so we are assuming parameter changes only occur in the z-direction. So although this special case has a better theoretical foundation than the forgetting strategy of [Caldas-Pinto 1983] and [Wagner 1987] it remains in essence a one-dimensional strategy. Furthermore we shall see from simulations that in general this method gives no practical advantage over the forgetting strategy of [Caldas-Pinto 1983] and [Wagner 1987].

4.5.2 Column forgetting—the basic algorithm

In §4.5.1 we showed that when $\lambda = 1$ two-dimensional forgetting could be implemented at far less computational cost than for the general case. However this algorithm has little merit, certainly over the more general one used in [Caldas-Pinto 1983] and [Wagner 1987]. Here we show that computation can again be saved when $\lambda = 0$. Furthermore this case may be more suitable for certain applications than the more general case.

Recall once again (Equation 4.2.6) that our estimate $\hat{\theta}(m,n)$ is given by

$$\hat{\theta}(m,n) = [Q_l(m,n) + Q_r(m,n)]^{-1}[R_l(m,n) + R_r(m,n)]$$

When $\lambda = 0$ we have the following relations.

$$Q_l(m,n) = \mu Q_l(m, n-1) + \phi(m,n)\phi^T(m,n)$$

$$Q_r(m,n) = 0$$

$$R_l(m,n) = \mu R_l(m, n-1) + \phi(m,n)y(m,n)$$

and

$$R_r(m,n) = 0$$

Thus as in the case with $\lambda = 1$ we may invoke the matrix inversion lemma for each update, but this time we update in the vertical direction. As a first example we may implement the following AML algorithm.

Algorithm 4.5.1 (Two-dimensional forgetting when $\lambda = 0$)

At each pixel (m,n),

(1) Form $\phi(m,n)$ using the new data and using the residuals $\eta(m-i, n-j)$ as estimates for the noise $e(m-i, n-j)$.

(2) Form the prediction error

$$\epsilon(m,n) = y(m,n) - \phi^T(m,n)\hat{\theta}(m, n-1)$$

and the residual

$$\eta(m,n) = \frac{\mu\epsilon(m,n)}{\mu + \phi^T(m,n)P(m, n-1)\phi(m,n)}$$

4.5 Row and column forgetting

(3) Update the new $P(m,n)$ according to

$$P(m,n) = \frac{P(m,n-1)}{\mu}\left[I - \frac{\phi(m,n)\phi^T(m,n)P(m,n-1)}{\mu + \phi^T(m,n)P(m,n-1)\phi(m,n)}\right]$$

(4) Form

$$\hat{\theta}(m,n) = \hat{\theta}(m,n-1) + P(m,n-1)\phi(m,n)\epsilon(m,n)$$

Notice that no more computation is required than for 2D-AML; however the storage requirement is greater since a whole row of P's and $\underline{\hat{\theta}}$'s must be stored at one time. The convergence rate will also be slow since only one column of information is used for each estimate.

For AR (and ARX) processes the algorithm is simply least squares but with a limited amount of data. At first sight this may seem a peculiar algorithm for ARMA models since the residuals are 'borrowed' from one estimator to another—ie $\phi(m,n)$ will include residuals $\eta(m-i, n-j)$'s with $i \neq 0$. However it is apparent that if the local supports are extended to include the whole width of the plane and $A(w^{-1}, z^{-1})$ and $C(w^{-1}, z^{-1})$ are both SWP-causal (see §3.4) then given n each $P(m,n)$ is identical. In other words in this case the algorithm is merely the multivariable AML algorithm described in [Borisson 1979]. Algorithm 4.5.1 may then be interpreted as multivariable AML but with the a priori assumptions that certain parameters are zero (since in general the local supports will not extend over the whole width of the plane).

4.5.3 Column forgetting—modifications

Our first interpretation of Algorithm 4.5.1 was as the two-dimensional forgetting algorithm but with λ set to zero. We have already noted in §4.1.1 that it is not necessary for any of our algorithms to run the full width of the plane. The immediate conclusion is that we can divide the plane into columns larger than the width of a pixel and run different estimators in each column. For example if we know we have the behaviour at the edges discussed in §3.3.7 then we could run separate estimators up the columns at the edges while running 2D-AML over the bulk of the data. This corresponds with running the two-dimensional forgetting factors with λ equal to zero at the edges but one away from the edges.

From a practical point of view, if we knew a priori that away from the edges the data really were stationary, it would save a lot of computation to replace the two-dimensional

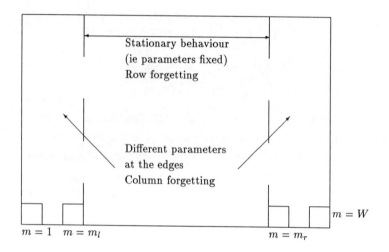

Fig 4.5.1 *Regions for Algorithm 4.5.2.*

forgetting away from the edges with the estimator used in [Caldas-Pinto 1983] and [Wagner 1987]. Suppose we wish to estimate the edges separately up to column m_l and beyond column m_r; our algorithm then becomes:

Algorithm 4.5.2 (Separate estimators over separate columns)

For each row n,

(1) For $m = 1$ to $m = m_l$,

    (i) Form $\phi(m, n)$ using the new data and using the residuals $\eta(m - i, n - j)$ as estimates for the noise $e(m - i, n - j)$.

    (ii) Form the prediction error

$$\epsilon(m, n) = y(m, n) - \phi^T(m, n)\hat{\theta}(m, n - 1)$$

and the residual

$$\eta(m, n) = \frac{\mu \epsilon(m, n)}{\mu + \phi^T(m, n) P(m, n - 1) \phi(m, n)}$$

4.5 Row and column forgetting

(iii) Update the new $P(m,n)$ according to

$$P(m,n) = \frac{P(m,n-1)}{\mu}\left[I - \frac{\phi(m,n)\phi^T(m,n)P(m,n-1)}{\mu + \phi^T(m,n)P(m,n-1)\phi(m,n)}\right]$$
(4.5.6)

(iv) Form

$$\hat{\theta}(m,n) = \hat{\theta}(m,n-1) + P(m,n-1)\phi(m,n)\epsilon(m,n)$$

(2) For $m = m_l + 1$ to $m = m_r - 1$,

(i) Form $\phi(m,n)$ using the new data and using the residuals $\eta(m-i, n-j)$ as estimates for the noise $e(m-i, n-j)$.

(ii) Form the prediction error

$$\epsilon(m,n) = \begin{cases} y(m,n) - \phi^T(m,n)\hat{\theta}(m-1,n) & \text{for } m > m_l + 1 \\ y(m,n) - \phi^T(m,n)\hat{\theta}(m_r - 1, n-1) & \text{for } m = m_l + 1 \end{cases}$$
(4.5.7)

and the residual

$$\eta(m,n) = \begin{cases} \dfrac{\mu\epsilon(m,n)}{\mu + \phi^T(m,n)P(m-1,n)\phi(m,n)} & \text{for } m > m_l + 1 \\ \dfrac{\mu\epsilon(m,n)}{\mu + \phi^T(m,n)P(m_r - 1, n-1)\phi(m,n)} & \text{for } m = m_l + 1 \end{cases}$$
(4.5.8)

(iii) Update the new $P(m,n)$ according to

$$P(m,n) = \begin{cases} \dfrac{P(m-1,n)}{\mu} \times \\ \left[I - \dfrac{\phi(m,n)\phi^T(m,n)P(m-1,n)}{\mu + \phi^T(m,n)P(m-1,n)\phi(m,n)}\right] \\ \qquad\qquad\qquad \text{for } m > m_l + 1 \\[6pt] \dfrac{P(m_r - 1, n-1)}{\mu} \times \\ \left[I - \dfrac{\phi(m,n)\phi^T(m,n)P(m_r - 1, n-1)}{\mu + \phi^T(m,n)P(m_r - 1, n-1)\phi(m,n)}\right] \\ \qquad\qquad\qquad \text{for } m = m_l + 1 \end{cases}$$
(4.5.9)

(iv) Form

$$\hat{\theta}(m,n) = \begin{cases} \hat{\theta}(m-1,n) \\ \quad + P(m-1,n)\phi(m,n)\epsilon(m,n) & \text{for } m > m_l + 1 \\ \hat{\theta}(m_r-1,n-1) \\ \quad + P(m_r-1,n-1)\phi(m,n)\epsilon(m,n) & \text{for } m = m_l + 1 \end{cases}$$
(4.5.10)

(3) For $m = m_r$ to $m = W$ repeat the steps of (1).

4.6 Simulations

In this section we compare the various forgetting algorithms. In the first series of tests we consider a two-parameter model given by

$$\left(1 + awz^{-1}\right) y(m,n) = \left(1 + cz^{-1}\right) e(m,n)$$

where a and c are allowed to change. The small number of parameters should ensure favourable conditions for the algorithms to react to parameter changes. The following algorithms were compared:

- General two-dimensional forgetting—the specific implementation used was Algorithm 4.4.2. Hence the algorithm was implemented as 2D-AML using Roesser's state-space form and with Givens rotations.

- Wagner's algorithm—a 2D-AML form of the original forgetting algorithm in [Wagner 1987]. Algorithm 4.1.1 was implemented but with $P(m,n)$ updated according to Equation 4.2.1 and including the residuals as estimates of the noise.

- Column forgetting—Algorithm 4.5.1.

- Row forgetting—Algorithm 4.1.1 was implemented but with $P(m,n)$ updated according to Equation 4.5.5 and including the residuals as estimates of the noise.

For most forgetting schemes there is a trade-off between convergence and response to changes in the parameters. To ensure a fair comparison between the four algorithms they were first implemented on the model

$$\left(1 - 0.8wz^{-1}\right) y(m,n) = \left(1 + 0.7z^{-1}\right) e(m,n) \tag{4.6.1}$$

4.6 Simulations

with no changes to the parameters. The process was generated over 100 by 300 pixels. In each case forgetting factors were chosen such that the mean error squared of the a estimate was approximately 0.003 when measured over the last 50 rows and between the 11th and 90th columns. These forgetting factors were then used in the subsequent tests.

Algorithm	Mean error squared for parameter estimates		Forgetting factors
	$\hat{a}$	$\hat{c}$	
General two-dimensional forgetting	0.0027	0.0132	$\lambda = \mu = 0.77$
Wagner's algorithm	0.0030	0.0067	$\lambda = 0.98$
Column forgetting	0.0027	0.0063	$\mu = 0.99$
Row forgetting	0.0030	0.0090	$\mu = 0.22$

Table 4.6.1 *Chosen values of forgetting factors for forgetting algorithms.*

The four algorithms were then tested on four different processes. These were again generated over 100 by 300 pixels using Equation 4.6.1. However this was allowed to change to the process

$$\left(1 + 0.8wz^{-1}\right) y(m,n) = \left(1 - 0.7z^{-1}\right) e(m,n) \qquad (4.6.2)$$

The boundary for the change was in four different directions, namely (i) horizontal, (ii) diagonal (see Fig 4.6.1), (iii) vertical and (iv) circular (see Fig 4.6.2).

Here (Fig's 4.6.3 to 4.6.14) we show the results of general two-dimensional forgetting and Wagner's algorithm when applied to cases (i) and (iii) (horizontal and vertical changes respectively). A full set of results is presented in Appendix 4. Not only are two-dimensional 'meshes' shown but also cross-sections as specified in Figs 4.6.1 and 4.6.2. The full set of 'meshes' are listed in Table 4.6.2 and the set of cross-sections in Table 4.6.3.

The first observation is that Wagner's algorithm and row forgetting behave very similarly for all the experiments. In these circumstances there is very little to choose between them as algorithmically they are also very similar. Heuristically they differ only in that Wagner's algorithm forgets a little at each update while row forgetting forgets only at the start of each row (when it forgets a great deal). One might then expect Wagner's algorithm to show a smoother (though not necessarily more accurate)

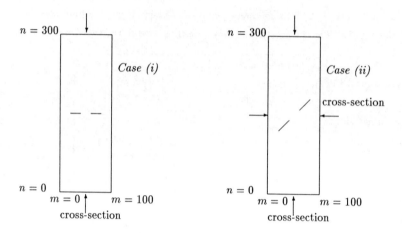

Fig 4.6.1 *Cases (i) and (ii)—horizontal and diagonal changes respectively.*

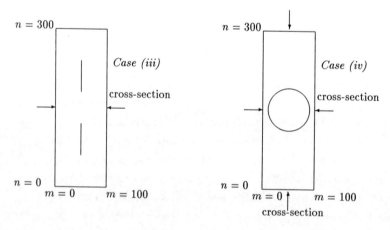

Fig 4.6.2 *Cases (iii) and (iv)—vertical and circular changes respectively.*

4.6 Simulations

Fig 4.6.3 *'Mesh' of â for case (i) using general two-dimensional forgetting.*

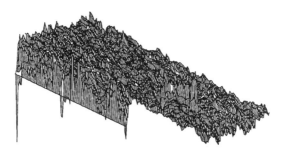

Fig 4.6.4 *'Mesh' of ĉ for case (i) using general two-dimensional forgetting.*

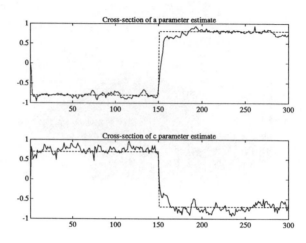

Fig 4.6.5 *Vertical cross-section of â and ĉ for case (i) using general two-dimensional forgetting.*

Fig 4.6.6 *'Mesh' of â for case (iii) using general two-dimensional forgetting.*

4.6 Simulations

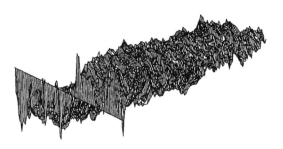

Fig 4.6.7 'Mesh' of $\hat{c}$ for case (iii) using general two-dimensional forgetting.

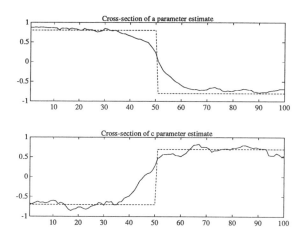

Fig 4.6.8 Horizontal cross-section of $\hat{a}$ and $\hat{c}$ for case (iii) using general two-dimensional forgetting.

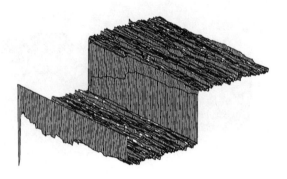

Fig 4.6.9 *'Mesh' of $\hat{a}$ for case (i) using Wagner's algorithm.*

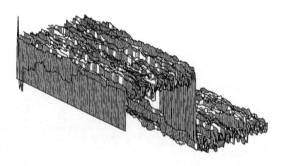

Fig 4.6.10 *'Mesh' of $\hat{c}$ for case (i) using Wagner's algorithm.*

4.6 Simulations

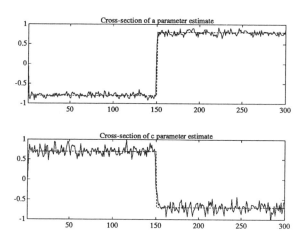

Fig 4.6.11 *Vertical cross-section of $\hat{a}$ and $\hat{c}$ for case (i) using Wagner's algorithm.*

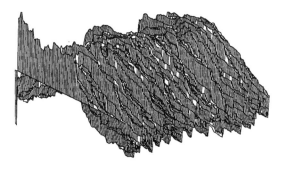

Fig 4.6.12 *'Mesh' of $\hat{a}$ for case (iii) using Wagner's algorithm.*

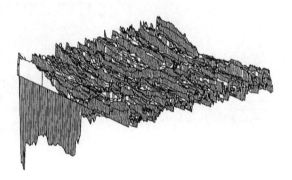

Fig 4.6.13 *'Mesh' of ĉ for case (iii) using Wagner's algorithm.*

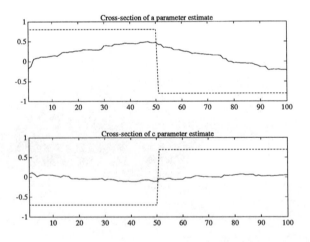

Fig 4.6.14 *Horizontal cross-section of â and ĉ for case (iii) using Wagner's algorithm.*

4.6 Simulations

		Horizontal	Diagonal	Vertical	Circular	Edges
General two-dimensional forgetting	vertical	Fig 4.6.5	Fig A4.3		Fig A4.7	
	horizontal		Fig A4.4	Fig 4.6.8	Fig A4.8	Fig 4.6.16
Wagner's algorithm	vertical	Fig 4.6.11	Fig A4.11		Fig A4.15	
	horizontal		Fig A4.12	Fig 4.6.14	Fig A4.16	
Column forgetting	vertical	Fig A4.19	Fig A4.22		Fig A4.29	
	horizontal		Fig A4.23	Fig A4.26	Fig A4.30	Fig 4.6.17
Row forgetting	vertical	Fig A4.33	Fig A4.36		Fig A4.44	
	horizontal		Fig A4.37	Fig A4.40	Fig A4.44	
Hybrid algorithm	vertical					
	horizontal					Fig 4.6.18

Table 4.6.3 *Figures for the cross-sections ($\hat{a}$ and $\hat{c}$ are shown on the same Figure).*

the familiar result that Wagner's algorithm works better than general two-dimensional forgetting when the change is horizontal but fails entirely when the change is vertical. However in all four cases the convergence is poor. To improve convergence for the two-dimensional forgetting the algorithm was implemented again but this time with $\lambda = 0.95$ and $\mu = 0.95$. The results for this case are shown in Figs 4.6.27 to 4.6.30. This time, although the estimates are much smoother, the response to change is slow. This suggests that, although two-dimensional forgetting has been shown to be more versatile than Wagner's algorithm when changes occur in different directions, it may be hard to find a good balance in the trade-off between estimate convergence and response to parameter change.

128   4 Parameter estimation

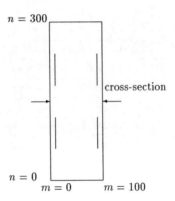

Fig 4.6.15 *Edge changes.*

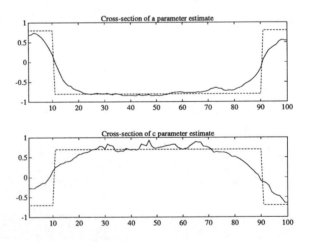

Fig 4.6.16 *Horizontal cross-section of $\hat{a}$ and $\hat{c}$ for case (v) using general two-dimensional forgetting.*

4.6 Simulations

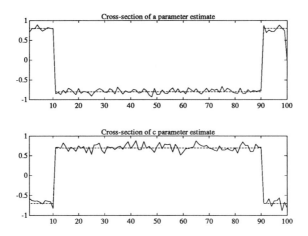

Fig 4.6.17 *Horizontal cross-section of $\hat{a}$ and $\hat{c}$ for case (v) using column forgetting.*

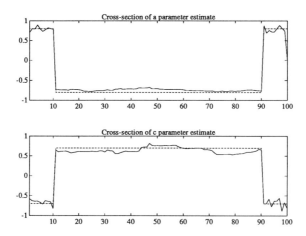

Fig 4.6.18 *Horizontal cross-section of $\hat{a}$ and $\hat{c}$ for case (v) using column forgetting at the edges.*

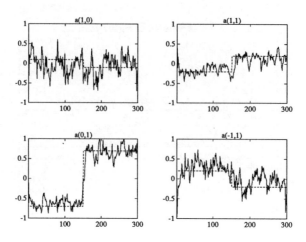

Fig 4.6.19 *Vertical cross-sections of $\hat{a}$ parameters for case (i) with eight parameters using general two-dimensional forgetting with λ and μ set to 0.77.*

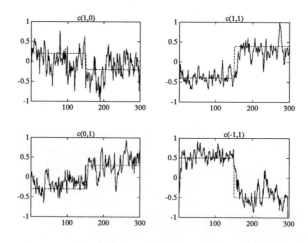

Fig 4.6.20 *Vertical cross-sections of $\hat{c}$ parameters for case (i) with eight parameters using general two-dimensional forgetting with λ and μ set to 0.77.*

4.6 Simulations

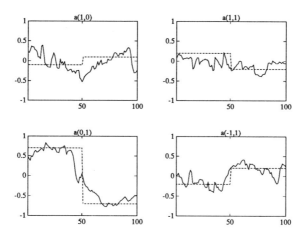

Fig 4.6.21 *Horizontal cross-sections of â parameters for case (iii) with eight parameters using general two-dimensional forgetting with λ and μ set to 0.77.*

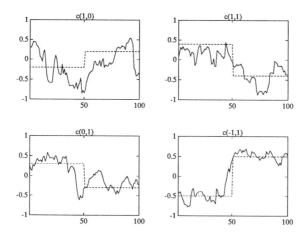

Fig 4.6.22 *Horizontal cross-sections of ĉ parameters for case (iii) with eight parameters using general two-dimensional forgetting with λ and μ set to 0.77.*

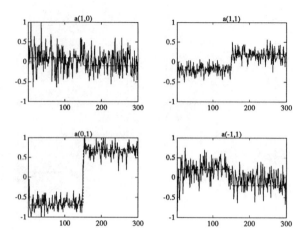

Fig 4.6.23 *Vertical cross-sections of $\hat{a}$ parameters for case (i) with eight parameters using Wagner's algorithm with λ set to 0.98.*

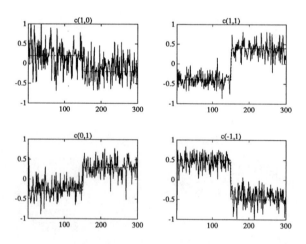

Fig 4.6.24 *Vertical cross-sections of $\hat{c}$ parameters for case (i) with eight parameters using Wagner's algorithm with λ set to 0.98.*

4.6 Simulations

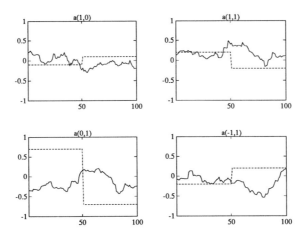

Fig 4.6.25 *Horizontal cross-sections of â parameters for case (iii) with eight parameters using Wagner's algorithm with λ set to 0.98.*

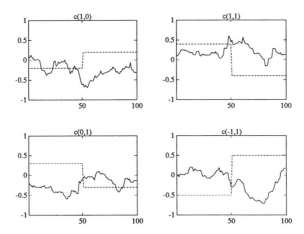

Fig 4.6.26 *Horizontal cross-sections of ĉ parameters for case (iii) with eight parameters using Wagner's algorithm with λ set to 0.98.*

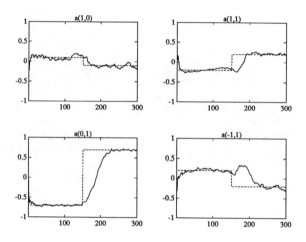

Fig 4.6.27 *Vertical cross-sections of â parameters for case (i) with eight parameters using general two-dimensional forgetting with λ and μ set to 0.95.*

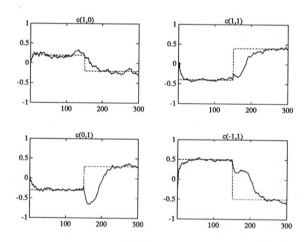

Fig 4.6.28 *Vertical cross-sections of ĉ parameters for case (i) with eight parameters using general two-dimensional forgetting with λ and μ set to 0.95.*

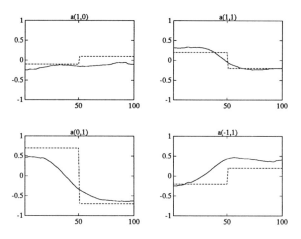

Fig 4.6.29 *Horizontal cross-sections of â parameters for case (iii) with eight parameters using general two-dimensional forgetting with λ and μ set to 0.95.*

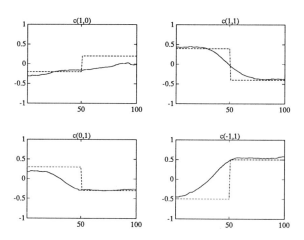

Fig 4.6.30 *Horizontal cross-sections of ĉ parameters for case (iii) with eight parameters using general two-dimensional forgetting with λ and μ set to 0.95.*

5 Self-tuning control

In §2 and §3 we developed prediction, minimum variance control and generalised minimum variance control algorithms for two-dimensional CARMA processes. In the spirit of the one-dimensional paradigms for our control algorithms [Astrom and Wittenmark 1973], [Clarke and Gawthrop 1975] and also the predictor for two-dimensional ARMA processes [Wagner and Wellstead 1990], [Wagner 1987] these algorithms are designed to be combined with recursive parameter estimators as discussed in §4 and hence implemented in self-tuning mode. In this chapter we present such a combined algorithm and demonstrate by simulation the usual neat properties.

We consider two further problems in this chapter: namely setpoint tracking and offset handling. While both may be solved easily with parameters fixed it is more natural to consider them as part of the self-tuning problem. We show that to analyse the setpoint problem correctly we must use the multivariable theory considered in §3.3 if we wish to ensure good tracking everywhere, and in particular at the edges. Two procedures are presented for achieving a given setpoint; however the first requires the manipulation of large matrices while the second (integral control) may cause stability problems in two dimensions. The offset problem turns out to be a fairly straightforward generalisation of the one-dimensional case; however, just as in one dimension, if we cannot model the process as incremental then we encounter computational difficulties in the parameter estimation.

The possibilities for control with parameter estimation are considered in §5.1, both for infinite and finite width data fields. Setpoint tracking for control is discussed in §5.2 and offset handling in §5.3. Throughout we will consider the generalised minimum variance controller only, taking the philosophy that the minimum variance controller is simply a special case and that given the minimum variance controller the least squares predictor may easily be derived.

5.1 Self-tuning GMV in two dimensions

5.1.1 An explicit self-tuning controller

We begin by combining the two-dimensional generalised minimum variance controller of §3.1.4 with 2D-AML discussed in §4.1.2. Our model is the usual CARMA model given by

$$A(w^{-1}, z^{-1})y(m,n) = z^{-\nu}B(w^{-1}, z^{-1})u(m,n) + C(w^{-1}, z^{-1})e(m,n) \quad (5.1.1)$$

with $A(w^{-1}, z^{-1})$, $B(w^{-1}, z^{-1})$ and $C(w^{-1}, z^{-1})$ truncated appropriately at the edges as in §3.1.1. As in §4 we can express this as the regression model

$$y(m,n) = \phi^T(m,n)\theta + e(m,n) \quad (5.1.2)$$

where ϕ is the data vector and θ is the parameter vector. Note that since we now have a CARMA model (as opposed to the ARMA model considered in §4) the two vectors will include past input values and the $B(w^{-1}, z^{-1})$ polynomial parameters respectively.

Algorithm 5.1.1 (Explicit two-dimensional self-tuning generalised minimum variance control)

At each 'pixel' (m,n):

(1) Form the data vector $\phi(m,n)$ corresponding to the regression model (Equation 5.1.2) using past values of output y and input u, and substituting the residuals $\eta(m-i, n-j)$ as estimates for the true noise process $e(m-i, n-j)$.

(2) Form the prediction error

$$\epsilon(m,n) = \begin{cases} y(m,n) - \phi^T(m,n)\hat{\theta}(W, n-1) & \text{for } m = 1 \\ y(m,n) - \phi^T(m,n)\hat{\theta}(m-1, n) & \text{for } m \geq 2 \end{cases}$$

(3) Update the covariance matrix

$$P(m,n) = \begin{cases} P(W, n-1)\left[I - \dfrac{\phi(m,n)\phi^T(m,n)P(W, n-1)}{1 + \phi^T(m,n)P(W, n-1)\phi(m,n)}\right] & \text{for } m = 1 \\ \\ P(m-1, n)\left[I - \dfrac{\phi(m,n)\phi^T(m,n)P(m-1, n)}{1 + \phi^T(m,n)P(m-1, n)\phi(m,n)}\right] & \text{for } m \geq 2 \end{cases}$$

(4) Update the parameter estimate

$$\hat{\theta}(m,n) = \begin{cases} \hat{\theta}(W, n-1) + P(m,n)\phi(m,n)\epsilon(m,n) & \text{for } m = 1 \\ \hat{\theta}(m-1, n) + P(m,n)\phi(m,n)\epsilon(m,n) & \text{for } m \geq 2 \end{cases}$$

and form the new residual

$$\eta(m,n) = y(m,n) - \phi^T(m,n)\hat{\theta}(m,n)$$

(5) Using the values of $\hat{\theta}(m,n)$ obtain the polynomials $X_{m,0,\nu}$, $Y_{m,0,\nu}$ and $Z_{m,0,\nu}$ in the model for the pseudo-output

$$\begin{aligned} \psi(m,n) &= P_m y(m,n) + z^{-\nu} Q_m u(m,n) \\ &= X_{m,0,\nu} y(m, n-\nu) + Y_{m,0,\nu} u(m, n-\nu) + Z_{m,0,\nu} e(m, n-\nu) \\ &\quad + F_{m,0,\nu} e(m,n) \end{aligned}$$

(See §3.1.2 for the derivation of the polynomials $X_{m,0,\nu}$, $Y_{m,0,\nu}$ and $Z_{m,0,\nu}$.)

(6) Apply the controller

$$X_{m,0,\nu} y(m,n) + Y_{m,0,\nu} u(m,n) + Z_{m,0,\nu} \eta(m,n) = 0 \tag{5.1.3}$$

Note that the estimate of the noise process used in the control law (Equation 5.1.3) is the residual $\eta(m,n)$; this saves us from estimating the noise process by using the process model itself as we did in §3 (Equation 3.1.6). The algorithm above is presented to demonstrate only the feasibility of combining the prediction or control strategies with parameter estimation routines; many important details have been ignored. In particular in practice some form of forgetting would be included and the covariance matrix P (or alternatively the information matrix P^{-1}) would be updated in factored form, as discussed in §4.

As a first simulation suppose we apply this algorithm to our usual model (Equation 5.1.1) with the following polynomial values:

$$A_0(w^{-1}, z^{-1}) = 1 - 0.6w^{-1}z^{-1} - 0.4z^{-1} - 0.5wz^{-1} \tag{5.1.4}$$

$$B_0(w^{-1}, z^{-1}) = 1 + 0.5w^{-1}z^{-1} + 0.7z^{-1} + 0.4wz^{-1} \tag{5.1.5}$$

5.1 Self-tuning GMV in two dimensions

$$C_0(w^{-1}, z^{-1}) = 1 + 0.2w^{-1}z^{-1} - 0.3z^{-1} + 0.4wz^{-1} \tag{5.1.6}$$

and with delay

$$\nu_0 = 1 \tag{5.1.7}$$

Note that this is the same model as that used in the third simulation of §3.2. Both $A(w^{-1}, z^{-1})$ and $B(w^{-1}, z^{-1})$ are inverse unstable but they share no common inverse unstable zero. Set

$$P(w^{-1}, z^{-1}) = Q(w^{-1}, z^{-1}) = 1$$

Fig 5.1.1 shows the output, firstly as a two-dimensional 'mesh', and secondly with rows concatenated. Fig 5.1.2 compares the cumulative squared output (with rows again concatenated) with the output for the same controller but with parameters known. The estimates for the parameters of $A(w^{-1}, z^{-1})$ are shown in Fig 5.1.3, those of $B(w^{-1}, z^{-1})$ in Fig 5.1.4 and those of $C(w^{-1}, z^{-1})$ in Fig 5.1.5. Recall (Equation 5.1.3) that the controller polynomials are given by $X_{m,0,\nu}$, $Y_{m,0,\nu}$ and $Z_{m,0,\nu}$ and from §3.1 that sufficiently far from the edges they have constant values for all m. The estimates of their parameters are shown in Figs 5.1.6, 5.1.7 and 5.1.8 respectively. Note that they are normalised so that the first term of $Y_{m,0,\nu}$ is 1.

At this point we should note that the analysis of parameter estimation for two-dimensional processes is far from complete. We find that some results from one-dimensional theory extend to the two-dimensional case staightforwardly while others do not. A good example is persistent excitation: just as for any least squares type problem where the solution $\hat{\theta}$ is given by

$$\hat{\theta} = \left[X^T X\right]^{-1} X^T Y$$

for a unique solution we require the covariance matrix $\left[X^T X\right]^{-1}$ to have full rank. In one dimension this property is related to the order of persistent excitation of the input (see for example [Soderstrom and Stoica 1989] for a discussion of persistent excitation in the one-dimensional case). While in two dimensions the rank of the covariance matrix still depends on the input, there is apparently no simple definition for the order of persistent excitation of the input. Hence we must consider the covariance matrix itself rather than the input to determine whether a unique solution to the parameter estimation may be obtained.

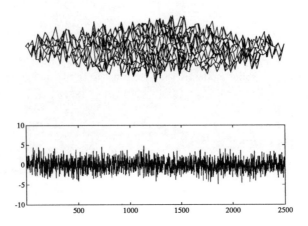

Fig 5.1.1 *Closed-loop output for the basic self-tuning two-dimensional generalised minimum variance controller.*

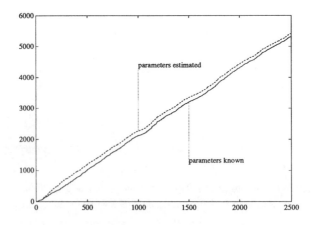

Fig 5.1.2 *Cumulative output squared for the generalised minimum variance controller.*

5.1 Self-tuning GMV in two dimensions

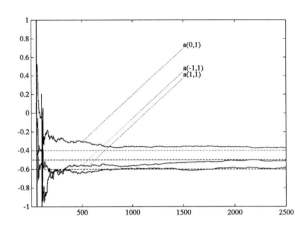

Fig 5.1.3 *Parameter estimates for $A(w^{-1}, z^{-1})$.*

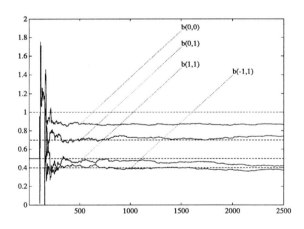

Fig 5.1.4 *Parameter estimates for $B(w^{-1}, z^{-1})$.*

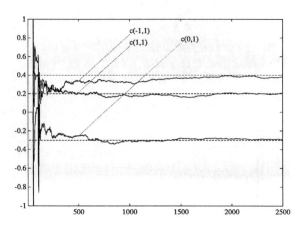

Fig 5.1.5 *Parameter estimates for $C(w^{-1}, z^{-1})$.*

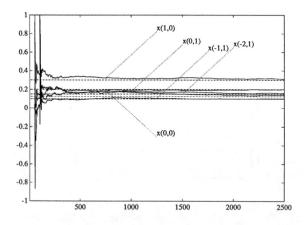

Fig 5.1.6 *Parameter estimates for the controller polynomial $X_{m,0,\nu}(w^{-1}, z^{-1})$ evaluated away from the edges.*

5.1 Self-tuning GMV in two dimensions

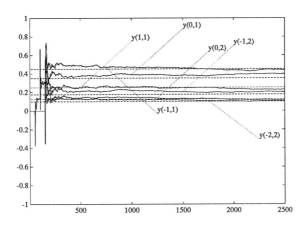

Fig 5.1.7 *Parameter estimates for the controller polynomial $Y_{m,0,\nu}(w^{-1}, z^{-1})$ evaluated away from the edges.*

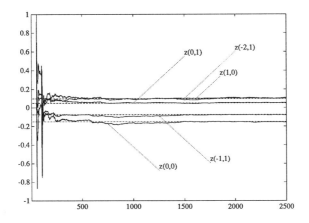

Fig 5.1.8 *Parameter estimates for the controller polynomial $Z_{m,0,\nu}(w^{-1}, z^{-1})$ evaluated away from the edges.*

It turns out that in the example above in steady state the covariance matrix will have only rank 9 while there are 10 parameters to estimate. The good behaviour of the controller may be explained heuristically as follows:

- The control law is expressed as a ratio of estimated parameters. Hence we have an extra degree of freedom in the derivation of the *controller* parameters. So in steady state the *controller* parameters will be correct (see Figs 5.1.6 to 5.1.8).

- When starting up the estimator the varying parameter estimates will cause a non-linear control law. Hence initially the covariance matrix will have full rank and the *model* parameter estimates will converge to values close to the true values (see Figs 5.1.3 to 5.1.5).

This phenomenon is similar to that familiar for one-dimensional explicit regulators where given certain conditions [Ljung and Soderstrom 1983] the *controller* parameters will converge to their true values despite linear dependence in the estimated model parameters. However the many convergence results for one-dimensional estimation have not yet been generalised to two dimensions (see [Wagner 1987] for a discussion). In this work we make no claims for the convergence properties of the algorithms. Note however that persistent excitation can always be achieved by increasing the complexity of the control law. Extending the controller in this way serves to break up the linear dependence in the regressors which causes the covariance matrix rank reduction. For the example, changing the $Q(w^{-1}, z^{-1})$ polynomial to

$$Q(w^{-1}, z^{-1}) = 1 + \lambda w^{-1} \text{ with } \lambda \neq 0$$

is sufficient.

5.1.2 Implicit self-tuning control

As a rule implicit schemes (where the parameters of the prediction or control law are estimated directly) are preferred to explicit schemes for one-dimensional processes since the computational burden is in general reduced. In [Wagner 1987] both explicit and implicit algorithms are developed for the prediction problem. However explicit algorithms are preferred since for the general two-dimensional problem explicit algorithms require fewer parameters to be estimated. Similarly it is straightforward to develop an implicit algorithm for generalised minimum variance control in the case where there are no edges to the data, viz:

5.1 Self-tuning GMV in two dimensions

Algorithm 5.1.2 (Implicit two-dimensional self-tuning generalised minimum variance control)

At each 'pixel' (m, n):

(1) Form pseudo-output $\psi(m, n)$ from

$$\psi(m, n) = Py(m, n) + z^{-\nu} Qu(m, n)$$

(2) Estimate $\hat{F}$ and $\hat{G}$ from the regression model

$$\psi(m, n) = \hat{F}u(m, n - \nu) + \hat{G}y(m, n - \nu) + \epsilon(m, n)$$

(3) Apply the controller

$$\hat{F}u(m, n) + \hat{G}y(m, n) = 0$$

The dimensions of $\hat{F}$ and $\hat{G}$ may be obtained by observing that their true values are given by

$$F = B\Phi' + [A(w^{-1}, 0)]^{\nu} QC$$

and

$$G = \Gamma'$$

as defined in §2.5.2.

The results of this controller when applied to the process defined by Equations 5.1.4 to 5.1.7 are shown in Fig 5.1.9. As for prediction in the two-dimensional case the algorithmic simplicity of an implicit self-tuner comes at the expense of more parameters to estimate. For the example the explicit algorithm requires ten parameters to be estimated while the implicit algorithm requires twelve. Furthermore we have a particularly conservative case in that

$$A(w^{-1}, 0) = 1$$

and ν is small. It is not hard to find cases where the difference in the number of parameters to estimate is much greater; for example if we take the same process as our example, but increase the delay to 2 the implicit controller would require twenty parameters to be estimated. For prediction it can be shown that the explicit predictor requires $2(2MN + M + N)$ parameters to be estimated while the implicit predictor requires

$$2MN + M + N + (N+1)M(2\nu+1) + (\mu+\nu M)N + \frac{MN}{2}(N+1)$$

parameters to be estimated. Since the algorithms are closely linked the orders are comparable for the corresponding generalised minimum variance control algorithms. Note that in one dimension implicit self-tuning algorithms also require more parameters than their explicit counterparts in general; it is the *order* of the increase in two dimensions which becomes prohibitive.

However, implicit controllers have the further disadvantage that edge conditions are hard to deal with. It turns out in our usual example that the implicit self-tuner works well (Fig 5.1.9). But suppose we apply both the explicit form and the implicit form of the generalised minimum variance controller to the process

$$A(w^{-1}, z^{-1})y(m,n) = z^{-1}B(w^{-1}, z^{-1})u(m,n) + C(w^{-1}, z^{-1})e(m,n) \tag{5.1.8}$$

with

$$A(w^{-1}, z^{-1}) = 1 - 0.7wz^{-1} - 1.1z^{-1} - 0.8w^{-1}z^{-1} \tag{5.1.9}$$

$$B(w^{-1}, z^{-1}) = 1 + 0.8wz^{-1} + 0.7z^{-1} + 0.9w^{-1}z^{-1} \tag{5.1.10}$$

and

$$C(w^{-1}, z^{-1}) = 1 + 0.2wz^{-1} - 0.4z^{-1} + 0.3w^{-1}z^{-1} \tag{5.1.11}$$

We can see from Fig 5.1.10 that not only does the implicit controller take longer to tune in, but even then it fails to control well at the edges. (Since the rows are concatenated the edges are represented by multiples of 50 along the x-axis. Thus the spikes that occur in Fig 5.1.10 represent poor control at the edges.) For these reasons we will limit our discussions below to explicit estimation algorithms.

5.1.3 Self-tuning control for different edge conditions

In §3.3.7 we noted that in practice we are likely to find richer behaviour near the edges than the simple assumption of §3.1.1. Specifically we might expect the parameters of the polynomials to have different values near the edges. Furthermore we noted that only simple changes were required for the generalised minimum variance controller to take into

5.1 Self-tuning GMV in two dimensions

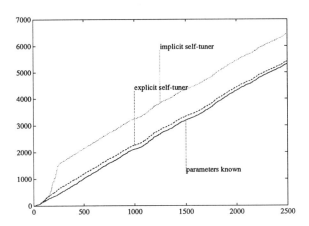

Fig 5.1.9 *Cumulative output squared comparing explicit and implicit controllers for the process given by Equations 5.1.4 to 5.1.7.*

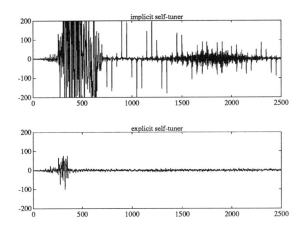

Fig 5.1.10 *Closed-loop output for the implicit and explicit controllers on the process given by Equations 5.1.8 to 5.1.11.*

account such changes provided we knew these different values. In §4.5.3 we presented an algorithm for estimating the parameters of such a model (Algorithm 4.5.2). Here we show the results of a simple example where the modified controller and estimation scheme are combined to form a self-tuning controller.

Fig 5.1.11 shows the cumulative output squared when three controllers were applied to the process given by

$$A_m(w^{-1}, z^{-1})y(m,n) = z^{-1}B_m(w^{-1}, z^{-1})u(m,n) + C_m(w^{-1}, z^{-1})e(m,n)$$

where

$$A_m(w^{-1}, z^{-1}) = \begin{cases} 1 - 0.3w^{-1}z^{-1} - 0.6z^{-1} - 0.2wz^{-1} \text{ for } m \leq 5 \\ \text{(truncated at the lefthand side where necessary)} \\ 1 - 0.6w^{-1}z^{-1} - 0.4z^{-1} - 0.5wz^{-1} \text{ for } 6 \leq m \leq 45 \\ 1 - 0.8w^{-1}z^{-1} - 0.4z^{-1} - 0.2wz^{-1} \text{ for } 46 \leq m \\ \text{(truncated at the righthand side where necessary)} \end{cases}$$

$$B_m(w^{-1}, z^{-1}) = \begin{cases} -0.3 - 0.4w^{-1}z^{-1} + 0.7z^{-1} + 0.3wz^{-1} \text{ for } m \leq 5 \\ \text{(truncated at the lefthand side where necessary)} \\ 1 + 0.5w^{-1}z^{-1} + 0.7z^{-1} + 0.4wz^{-1} \text{ for } 6 \leq m \leq 45 \\ 0.4 - 0.3w^{-1}z^{-1} + 0.2z^{-1} + 0.4wz^{-1} \text{ for } 46 \leq m \\ \text{(truncated at the righthand side where necessary)} \end{cases}$$

and

$$C_m(w^{-1}, z^{-1}) = \begin{cases} 1 + 0.2w^{-1}z^{-1} - 0.3z^{-1} - 0.4wz^{-1} \text{ for } m \leq 5 \\ \text{(truncated at the lefthand side where necessary)} \\ 1 + 0.2w^{-1}z^{-1} - 0.3z^{-1} + 0.4wz^{-1} \text{ for } 6 \leq m \leq 45 \\ 1 + 0.2w^{-1}z^{-1} - 0.3z^{-1} - 0.1wz^{-1} \text{ for } 46 \leq m \\ \text{(truncated at the righthand side where necessary)} \end{cases}$$

The three controllers were:

(1) A two-dimensional generalised minimum variance controller modified as suggested in §3.3.7 to take into account the edge behaviour (with the parameters known).

(2) The same controller as (1) but estimating the parameters using Algorithm 4.5.2.

(3) A two-dimensional generalised minimum variance controller with the (false) assumption that the local supports are merely truncated at the edges.

5.2 Setpoint tracking

It can be seen (Fig 5.1.11) that after an initial tuning-in period the self-tuning controller behaves in a similar manner to the optimal with known parameters. Both these controllers behave better than that which assumes a false model at the edges.

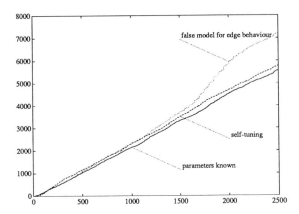

Fig 5.1.11 *Cumulative output squared for three controllers with a process whose behaviour changes at the edges.*

We saw in §4.6 that in general running separate parameter estimators up the columns will result in slower convergence. Hence we may expect a longer tuning-in phase at the edges for such a self-tuning controller. Note that in this case we could have employed the same estimator over all 5 of the lefthand columns and similarly over all 5 of the righthand columns since within these regions the parameters do not change. This emphasises the need to make the best use of all a priori knowledge about the process we are to control.

5.2 Setpoint tracking

5.2.1 Tracking without edges

In §2.5.2 and §3.1.4 we included a setpoint in our derivation of the two-dimensional generalised minimum variance controller. Recall (§2.5.2) that for the case with no edges the control law for the criterion

$$\min_{u(m,n)} E\left\{[Py(m, n+\nu) - Rr(m,n)]^2 + [Q'u(m,n)]^2\right\}$$

is

$$\left\{B\Phi' + [A(w^{-1},0)]^\nu QC\right\} u(m,n) + \Gamma' y(m,n) = [A(w^{-1},0)]^\nu r(m,n)$$

for certain polynomials Φ' and Γ' giving closed-loop output

$$y(m,n) = \frac{z^{-\nu}BR}{AQ+BP} r(m,n) + \frac{B\Phi' + [A(w^{-1},0)]^\nu QC}{[A(w^{-1},0)]^\nu (AQ+BP)} e(m,n)$$

We can employ the final-value theorem to say

$$\lim_{n\to\infty} E[y(m,n)] = \frac{B(w^{-1},1)R(w^{-1},1)}{A(w^{-1},1)Q(w^{-1},1) + B(w^{-1},1)P(w^{-1},1)} r(m,n)$$

So to achieve the correct setpoint we require

$$\frac{B(1,1)R(1,1)}{A(1,1)Q(1,1) + B(1,1)P(1,1)} = 1 \tag{5.2.1}$$

The simplest way to ensure this is to set

$$R(w^{-1},z^{-1}) = \frac{A(1,1)Q(1,1) + B(1,1)P(1,1)}{B(1,1)} \tag{5.2.2}$$

provided

$$B(1,1) \neq 0$$

Suppose we apply such a criterion to the case with edges. To achieve the setpoint this requires us to be sufficiently far from the lefthand edge (setting w^{-1} to 1 in Equation 5.2.1 is a steady-state argument in the *horizontal* direction). Furthermore the argument for its derivation assumes there is no righthand edge. It follows that such a strategy should fail at the edges.

Suppose we were to apply the technique to the model given by Equations 5.1.4 to 5.1.7 and with

$$P(w^{-1},z^{-1}) = Q(w^{-1},z^{-1}) = 1$$

Then Equation 5.2.2 gives

$$R(w^{-1},z^{-1}) = \frac{(1 - 0.5 - 0.4 - 0.6) + (1 + 0.4 + 0.7 + 0.5)}{(1 + 0.4 + 0.7 + 0.5)}$$

$$= 0.8077 \text{ to 4 significant figures}$$

5.2 Setpoint tracking

Fig 5.2.1 shows both a 'mesh' of the output in this case where the setpoint is 100 and also a profile of the last row. (Subsequent plots of simulations have the same format.) It can be seen that the correct setpoint is not attained at the edges. However the fit is perhaps surprisingly good away from the edges.

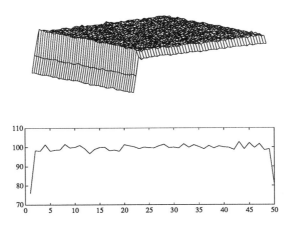

Fig 5.2.1 *Output with setpoint 100 with scalar weighting.*

5.2.2 Tracking column by column

To achieve good tracking in all columns we must use the multivariable analysis of §3.3. There are then two options for ensuring the correct setpoint is achieved everywhere. The first, which we consider in this section, requires the manipulation of scalar matrices corresponding to the multivariable problem. The second option is to use integral control which we consider in §5.2.3 (below).

Recall (Equation 3.3.7) that the closed-loop output for generalised minimum variance control is given by

$$\left(\bar{\underline{Q}}\,\underline{\underline{B}}^{-1}\underline{\underline{A}} + \underline{\underline{P}}\right)\underline{y} = z^{-\nu}\underline{\underline{R}}\,\underline{r} + \left(\bar{\underline{Q}}\,\underline{\underline{B}}^{-1}\underline{\underline{C}} + \underline{\underline{F}}\right)\underline{e}$$

Then suppose we wish to have steady-state output

$$\underline{y}(n) = \underline{y}_0$$

then we require

$$\left(\bar{\underline{Q}}(1) \left[\underline{\underline{B}}(1)\right]^{-1} \underline{\underline{A}}(1) + \underline{\underline{P}}(1)\right) \underline{y}_0 = \underline{\underline{R}}(1)\underline{r} \qquad (5.2.3)$$

One way to ensure this condition is satisfied is to choose

$R(w^{-1}, z^{-1}) = 1$ and hence $\underline{\underline{R}}(z^{-1}) = \underline{\underline{I}}$

and then

$$\underline{r} = \left(\bar{\underline{Q}}(1) \left[\underline{\underline{B}}(1)\right]^{-1} \underline{\underline{A}}(1) + \underline{\underline{P}}(1)\right) \underline{y}_0 \qquad (5.2.4)$$

Fig 5.2.2 shows both a mesh of the output for this case again with a setpoint of 100 and also a profile of the last row. It can be seen that this time the setpoint is attained everywhere. Fig 5.2.3 shows the corresponding input. Fig 5.2.4 shows the values of $\underline{r}$ (scaled over 100). To the eye the values appear as a straight line away from the edges, with the same value (0.8077 to 4 significant figures) as we derived for $R(w^{-1}, z^{-1})$ in the previous case without edges.

It is clear that the disadvantage of this technique is the necessity of inverting large matrices; in particular this would be computationally very arduous if performed at each step in a self-tuning algorithm. However, this result suggests that it should be possible to use the scalar value everywhere except at the edges, so the full inverse need not be calculated. Alternatively the sparseness and symmetry of the matrices in Equation 5.2.4 can be exploited to reduce the computation considerably. In particular once again we can borrow from the analysis of partial differential equations [Richtmyer 1957] the same technique we exploited in §3.4 to find the solution without explicitly inverting $\underline{\underline{B}}$. Specifically:

Suppose we wish to find $\underline{\xi}$ where

$$\underline{\xi} = \left[\underline{\underline{B}}(1)\right]^{-1} \underline{\chi}$$

for some known $\underline{\chi}$. Then

$$\sum_{i=-M}^{M} \beta_i \xi(m-i) = \chi(m)$$

where β_i is the $(m, m-i)$'th entry of $\underline{\underline{B}}(1)$ and

$$\xi(m-i) = 0 \text{ for } \begin{cases} m-i < 1 \\ m-i > W \end{cases}$$

5.2 Setpoint tracking

Then in order of increasing m we may find new $\beta'_{m,i}$'s and $\chi'(m)$ such that

$$\sum_{i=-M}^{0} \beta'_{m,i}\xi(m-i) = \chi'(m)$$

and then in order of *decreasing* m we may solve for $\xi(m)$.

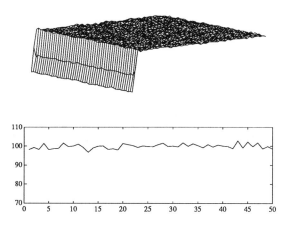

Fig 5.2.2 *Output with setpoint* 100 *with each column weighted separately.*

5.2.3 Integral control

The alternative to weighting each column separately is to use integral control. Just as in one dimension, we choose $\underline{\underline{Q}}(z^{-1})$ to have a factor $(1-z^{-1})$ so that

$$\underline{\underline{Q}}(1) = 0$$

Then Equation 5.2.3 becomes simply

$$\underline{\underline{P}}(1)\underline{y}_0 = \underline{\underline{R}}(1)\underline{r}$$

So to achieve the correct setpoint we need simply allow $\underline{\underline{Q}}(z^{-1})$ to have a factor $(1-z^{-1})$ and set

$$\underline{\underline{R}}(z^{-1}) = \underline{\underline{P}}(z^{-1})$$

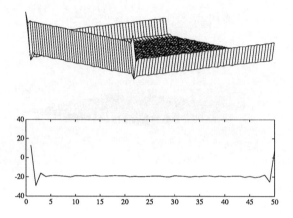

Fig 5.2.3 *Input with setpoint* 100 *with each column weighted separately.*

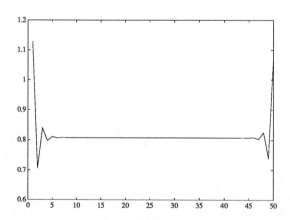

Fig 5.2.4 *The weightings for each column.*

5.2 Setpoint tracking

But

$$(1 - z^{-1}) \text{ divides } \underline{\underline{Q}}(z^{-1}) \Leftrightarrow (1 - z^{-1}) \text{ divides } Q(w^{-1}, z^{-1})$$

and

$$\underline{\underline{R}}(z^{-1}) = \underline{\underline{P}}(z^{-1}) \Leftrightarrow R(w^{-1}, z^{-1}) = P(w^{-1}, z^{-1})$$

So we need simply allow $Q(w^{-1}, z^{-1})$ to have a factor $(1 - z^{-1})$ and set

$$R(w^{-1}, z^{-1}) = P(w^{-1}, z^{-1})$$

Fig 5.2.5 shows the results for the our usual algorithm with

$$P(w^{-1}, z^{-1}) = R(w^{-1}, z^{-1}) = 1$$

and

$$Q(w^{-1}, z^{-1}) = 1 - z^{-1}$$

Fig 5.2.6 shows the corresponding input. It can be seen that the settling time is longer than for the previous case but that the correct setpoint is achieved everywhere.

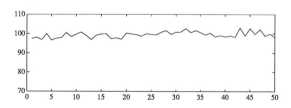

Fig 5.2.5 *Output with setpoint* 100 *with integral control.*

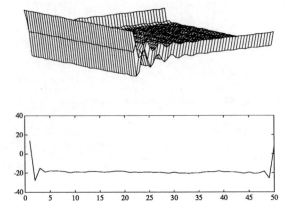

Fig 5.2.6 *Input with setpoint* 100 *with integral control.*

At first sight it seems that introducing integral action provides a simple way of achieving a given setpoint. But in two dimensions ensuring stability of the denominator of the closed-loop output for the generalised minimum variance controller is not necessarily easy. And whereas before we had only to consider the stability of

$$P(w^{-1}, z^{-1})B(w^{-1}, z^{-1}) + Q_1(w^{-1}, z^{-1})A(w^{-1}, z^{-1})$$

we must now consider

$$P(w^{-1}, z^{-1})B(w^{-1}, z^{-1}) + (1 - z^{-1})Q_2(w^{-1}, z^{-1})A(w^{-1}, z^{-1})$$

for some Q_1 and Q_2. In particular if $B(w^{-1}, 1)$ has an unstable zero at

$$w^{-1} = b$$

then integral action cannot stabilise the closed-loop output, since the denominator of the closed-loop output has a zero at

$$(w^{-1}, z^{-1}) = (b, 1)$$

and the closed-loop response is at best marginally stable (see §2.3).

As an interesting case consider the effect of integral action when B is symmetric. We may say:

5.2 Setpoint tracking

Theorem 5.2.1 *If $B(w^{-1}, z^{-1})$ is symmetric then we can introduce integral control of the form $(1 - z^{-1})$ if and only if all factors of*

$$B|_{z^{-1}=1} = B_0 \prod_i (1 + \beta_i w^{-1} + \beta_i w)$$

have

$$|\beta| < 0.5$$

Proof *See Appendix 1.*

Consider some examples:

(1) With $B(w^{-1}, z^{-1})$ inverse stable we may always introduce integral control. Suppose for example

$$B(w^{-1}, z^{-1}) = 1 + \frac{1}{4}w^{-1}z^{-1} + \frac{1}{4}wz^{-1}$$

Then

$$B|_{z^{-1}=1} = 1 + \frac{1}{4}w^{-1} + \frac{1}{4}w$$

(2) With $B(w^{-1}, z^{-1})$ inverse unstable we may or may not be able to introduce integral control. For example if

$$B(w^{-1}, z^{-1}) = 1 + w^{-1}z^{-1} + wz^{-1}$$

Then

$$B|_{z^{-1}=1} = 1 + w^{-1} + w$$

so integral control fails, but if

$$B(w^{-1}, z^{-1}) = 1 + 5z^{-1} + w^{-1}z^{-1} + wz^{-1}$$

Then

$$B|_{z^{-1}=1} = 6 + w^{-1} + w$$

and it is possible to incorporate integral control.

To summarise, integral action provides a simple way of ensuring the correct setpoint is achieved, without involving excessive computation. However its introduction makes it harder to choose polynomials P and Q to ensure closed-loop stability; furthermore in certain circumstances its introduction may make it *impossible* to choose polynomials P and Q to ensure closed-loop stability.

It is worth noting at this stage that in two dimensions integral control does not necessarily appear in the form $(1 - z^{-1})$. It is clear that any polynomial $T(w^{-1}, z^{-1})$ such that

$$T|_{z^{-1}=1} = 0$$

will have similar effects, although behaviour may not be so straightforward at the edges. Furthermore we may also introduce integral action in the horizontal direction by incorporating any polynomial which is zero when $w^{-1} = 1$.

5.3 Offset handling

So far we have not considered the problem of offset and have allowed our disturbance data $e(m, n)$ to have zero mean. In any industrial process which we may wish to model as two-dimensional there is likely to be some non-zero offset. For example in papermaking if no cross-machine direction basis weight control is applied there is an uneven profile across the web. It is the steady state of this profile that current cross-machine direction control algorithms attempt to flatten [Wilhelm and Fjeld 1983]. We will consider two ways of introducing offset to our model. The first we call the incremental model given by

$$A(w^{-1}, z^{-1})y(m,n) = z^{-\nu}B(w^{-1}, z^{-1})u(m,n) + C(w^{-1}, z^{-1})\frac{e(m,n)}{\Delta} \qquad (5.3.1)$$

where

$$\Delta = 1 - z^{-1}$$

The second we call the static model given by

$$A(w^{-1}, z^{-1})y(m,n) = z^{-\nu}B(w^{-1}, z^{-1})u(m,n) + C(w^{-1}, z^{-1})e(m,n) + d(m)$$

$$(5.3.2)$$

for some non-zero $d(m)$. Note that in each case the offset is essentially determined by the column—in the first instance since Δ operates entirely in the vertical direction and

5.3 Offset handling

in the second case by specifying $d = d(m)$. This seems to fit well with the behaviour of processes such as papermaking, but of course other schemes are possible. As in one dimension, the incremental model allows a neat solution to the problem; however there may be circumstances where it is an unsuitable model, in particular when the noise is purely sensor noise. For the static model we will see that the control problem is closely related to the setpoint problem of §5.2. However the static model introduces severe difficulties to the parameter estimation.

5.3.1 The incremental model

Our solution for the incremental model is a generalisation of the one-dimensional controller in [Tuffs and Clarke 1985]. However it differs in that we wish to derive an explicit controller. We will first consider the estimation of parameters. Our model (Equation 5.3.1) may be written in regression form as

$$\Delta y(m,n) = A'(w^{-1}, z^{-1})(\Delta y(m-1,n)) + z^{-\nu} B(w^{-1}, z^{-1})(\Delta u(m,n))$$
$$+ C'(w^{-1}, z^{-1})e(m-1,n) + e(m,n)$$

$$= \phi^T(m,n)\theta(m,n) + e(m,n) \qquad (5.3.3)$$

where

$$w^{-1} A'(w^{-1}, z^{-1}) = 1 - A(w^{-1}, z^{-1})$$

and similarly

$$w^{-1} C'(w^{-1}, z^{-1}) = C(w^{-1}, z^{-1}) - 1$$

Thus we may implement a suitable estimation algorithm to obtain the parameters of A, B and C. Note that the data in the regression vector ϕ will have zero mean as recommended in [Clarke et al. 1983]. We may use the residuals in the estimation algorithm as estimates of the zero mean noise process $e(m,n)$. We may also find an estimate $\hat{x}$ for the offset term e/Δ recursively as

$$C(w^{-1}, z^{-1})\hat{x}(m,n) = A(w^{-1}, z^{-1})y(m,n) + z^{-\nu} B(w^{-1}, z^{-1})u(m,n) \qquad (5.3.4)$$

In algebraic terms the controller is constructed in a similar fashion to its one-dimensional counterpart. We may express the pseudo-output

$$\psi(m, n+\nu) = P(w^{-1}, z^{-1})y(m, n+\nu) + Q(w^{-1}, z^{-1})u(m,n)$$
$$- R(w^{-1}, z^{-1})r(m,n)$$

recursively as

$$A\psi(m, n+\nu) = [PB + QA]u(m,n) - ARr(m,n) + \frac{CP}{\Delta}e(m, n+\nu)$$

If we partition

$$\frac{CP}{\Delta} = AF' + z^{-\nu}\frac{G'}{\Delta}$$

where the support of $F'(w^{-1}, z^{-1})$ contains only future data then we may say

$$\begin{aligned}A\psi(m, n+\nu) &= [PB + QA]u(m,n) - ARr(m,n) \\ &\quad + \frac{G'}{\Delta}e(m,n) + AF'e(m, n+\nu) \\ &= \left[\frac{ABF'\Delta}{C} + QA\right]u(m,n) + \frac{AG'}{C}y(m,n) - ARr(m,n) \\ &\quad + AFe(m, n+\nu)\end{aligned}$$

Note that we have the usual result (§2.4 and §2.5) that while the polynomials $F'(w^{-1}, z^{-1})$ and $G'(w^{-1}, z^{-1})$ may have infinite support both

$$\Phi'(w^{-1}, z^{-1}) = [A(w^{-1}, 0)]^\nu F'(w^{-1}, z^{-1})$$

and

$$\Gamma'(w^{-1}, z^{-1}) = [A(w^{-1}, 0)]^\nu G'(w^{-1}, z^{-1})$$

have finite support.

The control law is then to set

$$[BF'\Delta + QC]u(m,n) + G'y(m,n) = RCr(m,n)$$

The closed-loop output is

$$[BP + QA]y(m,n) = z^{-\nu}BRr(m,n) + \frac{[BF'\Delta + QC]}{\Delta}e(m,n)$$

It follows that in order to ensure the setpoint is obtained (even if it is zero) we must choose

$$Q(w^{-1}, z^{-1}) = \Delta Q'(w^{-1}, z^{-1})$$

for some $Q'(w^{-1}, z^{-1})$. In other words we *must* use integral control in this case. Following the discussion in §5.2 we may conclude that we cannot achieve good control if $B(w^{-1}, 1)$ has an unstable zero.

5.3 Offset handling

Assuming B is such that we can use integral control, the closed-loop output is then

$$[BP + \Delta Q'C]\, y(m,n) = z^{-\nu} B R r(m,n) + [BF' + Q'C]\, e(m,n)$$

The correct setpoint is achieved provided R is chosen such that

$$R(w^{-1}, 1) = P(w^{-1}, 1)$$

As discussed before we cannot use this algebraic approach when we have finite edges, although it remains a useful analysis tool. The correct controller is easily derived following §3.1.4 from the definition of the pseudo-output

$$\begin{aligned}\psi(m, n+\nu) &= P(w^{-1}, z^{-1}) y(m, n+\nu) + \Delta Q'(w^{-1}, z^{-1}) u(m, n) \\ &\quad - R(w^{-1}, z^{-1}) r(m, n)\end{aligned}$$

and by repeatedly substituting for future values of y using the relationship

$$\begin{aligned}y(m+k_1, n+k_2) &= y(m+k_1, n+k_2-1) + (1-A)(\Delta y(m+k_1, n+k_2)) \\ &\quad + B(\Delta u(m+k_1, n+k_2-\nu)) + Ce(m+k_1, n+k_2)\end{aligned}$$

Alternatively we may substitute using the relationship

$$\begin{aligned}y(m+k_1, n+k_2) &= (1-A) y(m+k_1, n+k_2) + Bu(m+k_1, n+k_2-\nu) \\ &\quad + Cx(m+k_1, n+k_2)\end{aligned}$$

In this case we will need the estimates $\hat{x}$ obtained in Equation 5.3.4. We will also need the relation

$$x(m+k_1, n+k_2) = e(m+k_1, n+k_2) + x(m+k_1, n+k_2-1)$$

in order to partition future noise correctly.

To demonstrate this case consider the process given by

$$A_0 y(m,n) = z^{-1} B_0 u(m,n) + C_0 \frac{e(m,n)}{\Delta}$$

with A_0, B_0 and C_0 defined by Equations 5.1.4 to 5.1.6. Fig 5.3.1 shows the output when the generalised minimum variance controller is applied with parameters known and

$$P(w^{-1}, z^{-1}) = R(w^{-1}, z^{-1}) = 1$$

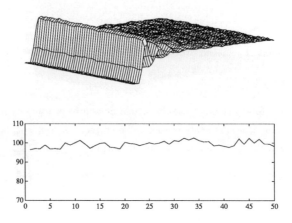

Fig 5.3.1 *Output with setpoint* 100 *for the incremental model, parameters known.*

and

$$Q(w^{-1}, z^{-1}) = 1 - z^{-1}$$

Fig 5.3.2 shows the corresponding input. It can be seen that the inclusion of incremental noise causes the input to lose its smoothness away from the edges.

Fig 5.3.3 shows the output for the same controller but in self-tuning form when parameters are unknown and so estimated.

5.3.2 The static model

There are occasions where the incremental model (Equation 5.3.1) may be inappropriate. In particular if the offset is constant (or varying at least as slowly as the parameters) and the remaining disturbance is stationary (for example most sensor noise without drift) then we must use the static model (Equation 5.3.2).

Control for this case is simple and best viewed as an extension of the setpoint problem considered in §5.2. However we shall see below that parameter estimation in this case is difficult.

5.3 Offset handling

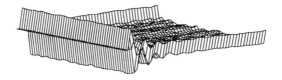

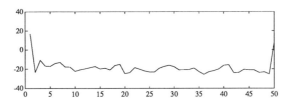

Fig 5.3.2 *Input with setpoint* 100 *for the incremental model, parameters known.*

Suppose we apply *any* controller of the form

$$X(w^{-1}, z^{-1})u(m,n) + Y(w^{-1}, z^{-1})y(m,n) = Z(w^{-1}, z^{-1})r(m,n)$$

Then the closed-loop output is

$$[AX + z^{-\nu}BY]\,y(m,n) = z^{-\nu}BZr(m,n) + CXe(m,n) + Xd(m)$$

Immediately it follows that if we use integral control, ie if

$$X(w^{-1}, 1) = 0$$

then the problem reduces to the integral control problem of §5.2.3. In this case no estimation of $d(m)$ is necessary for the control action.

Alternatively, if integral control is inappropriate, then as in §5.2.2 we must consider the problem in multivariable terms. This time our model becomes

$$\underline{\underline{A}}(z^{-1})\underline{y}(n) = z^{-\nu}\underline{\underline{B}}(z^{-1})\underline{u}(n) + \underline{\underline{C}}(z^{-1})\underline{e}(n) + \underline{d}$$

and our controller

$$\underline{\underline{X}}(z^{-1})\underline{u}(n) + \underline{\underline{Y}}(z^{-1})\underline{y}(n) = \underline{\underline{Z}}(z^{-1})\underline{r}(n)$$

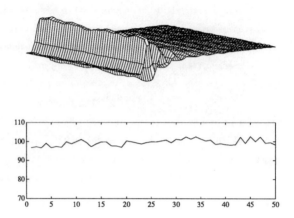

Fig 5.3.3 *Output with setpoint 100 for the incremental model, parameters estimated.*

The closed-loop output is

$$\left[\underline{\underline{X}}\,\underline{\underline{B}}^{-1}\underline{\underline{A}} + z^{-\nu}\underline{\underline{Y}}\right]\underline{y} = z^{-\nu}\underline{\underline{Z}}\,\underline{r} + \underline{\underline{X}}\,\underline{\underline{B}}^{-1}\underline{d}$$

Suppose our desired output is $\underline{y}_0$. Then we must choose $\underline{r}$ such that

$$\underline{\underline{Z}}(1)\underline{r} = \left[\underline{\underline{X}}(1)\underline{\underline{B}}^{-1}(1)\underline{\underline{A}}(1) + \underline{\underline{Y}}(1)\right]\underline{y}_0 - \underline{\underline{X}}(1)\underline{\underline{B}}^{-1}(1)\underline{d} \qquad (5.3.5)$$

It is clear that given an estimate of $\underline{d}$ such a controller is merely a slight modification of that in §5.2.2.

In this second case we have allowed $u(m,n)$ to be dependent explicitly on the offset (or an estimate of the offset in the self-tuning case). An alternative way of formulating the problem in this case is to reconsider the pseudo-output of two-dimensional generalised minimum variance control. As usual the pseudo-output is given by

$$\begin{aligned}\psi(m,n) &= P(w^{-1},z^{-1})y(m,n) + z^{-\nu}Q(w^{-1},z^{-1})u(m,n) \\ &\quad - z^{-\nu}R(w^{-1},z^{-1})r(m,n)\end{aligned}$$

In algebraic terms it is easy to show that the control law becomes

$$[BF + QC]\,u(m,n) + Fd(m) - RCr(m) + Gy(m,n) = 0$$

5.3 Offset handling

with closed-loop output

$$[BP + QA]\, y(m,n) = [BP + QC]\, e(m,n) + BRr(m,n) + Qd(m)$$

where we have partitioned

$$PC = AF + z^{-\nu}G$$

in the usual way, and must adjust these results suitably to allow for the edge conditions. The condition corresponding to Equation 5.3.5 is then to choose $r(m)$ such that

$$\left.\frac{BR}{BP+QA}\right|_{z^{-1}=1} r(m) + \left.\frac{Q}{BP+QA}\right|_{z^{-1}=1} d(m) = y_0(m)$$

where $y_0(m)$ is the desired steady-state output (note again that this is only an indication of the true condition as we do not consider edge conditions here). It is easy to see that an easy way to achieve this is to choose

$$Q(w^{-1}, z^{-1}) = (1 - z^{-1})Q'(w^{-1}, z^{-1})$$

for some Q' and with $r(m)$ equal to the desired output to set

$$R(w^{-1}, 1) = P(w^{-1}, 1)$$

Unfortunately, parameter estimation becomes problematic for the static model. In particular, despite the use of integral control, meaning an estimate of d is not required for the actual control law, it is still necessary to make such an estimate in order to correctly estimate the other parameters. There are then two classes of problem:

Firstly the problems encountered in the equivalent one-dimensional case apply here also. Specifically it is no longer appropriate to use a regression model of the form of Equation 5.3.3 since the noise process is no longer white gaussian. The alternative is to use the 'one-in-the-data' method. However this has the associated problems of both slowness of estimation and the covariance matrix becoming ill-conditioned [Clarke et al. 1983].

Secondly we have far more d parameters to estimate: in the one-dimensional case there is just one parameter whereas in two dimensions the number of extra parameters is equal to the width of the web. If

$$\phi_0^T = [y(m-1, n), \ldots, \eta(m - M, n - N)]$$

is the data vector before augmentation and ϕ_1 is the augmented data vector then

$$\phi_1'(m, n) = [y(m-1, n), \ldots, \eta(m - M, n - N), 0, 0, \ldots, 1, \ldots, 0]$$

with W extra terms all zero except for the m'th.

This enormous increase in the dimensions of the data vector, parameter vector and covariance matrix not only increases the amount of computation required but also slows down the convergence rate of the estimation. Note however that experiments show that the order of the condition number of the covariance matrix is determined by the greatest offset value; the increase in dimension associated with the two-dimensional problem should not affect the condition number very much.

One possibility for reducing the number of parameters to estimate is to use prior knowledge. For example if the offset is sufficiently smooth it may be reasonable to approximate it as some low-order polynomial or sum of sine functions. This is best illustrated by example. Consider the process given by

$$A_0(w^{-1}, z^{-1})y(m,n) = z^{-1}B_0(w^{-1}, z^{-1})u(m,n) + C_0(w^{-1}, z^{-1})e(m,n) + d(m)$$

with A_0, B_0 and C_0 defined by Equations 5.1.4 to 5.1.6 and $d(m)$ given by some fourth-order polynomial. Fig 5.3.4 shows the output when the generalised minimum variance controller described above is applied with parameters known. Fig 5.3.5 shows the corresponding input (since the offset is regulated out it is here that its effects may be observed). Fig 5.3.6 shows the corresponding output when the controller is applied in self-tuning mode. Here the data vector ϕ includes only five terms corresponding to the offset, and is given by

$$\phi(m,n) = \left[\cdots, \text{leg}_0(\frac{2m-51}{49}), \cdots, \text{leg}_4(\frac{2m-51}{49})\right]$$

Note that

$$1 \leq m \leq 50$$

and hence

$$-1 \leq \frac{2m-51}{49} \leq 1$$

Here $\text{leg}_i(x)$ is the Legendre polynomial of order i; these polynomials are suitable candidates[1] as entries in the data vector for two reasons. Firstly they are orthogonal over the interval

$$-1 \leq x \leq 1$$

---

[1] More recent work [Heath 1993] has shown the Legendre polynomials to be *inappropriate* candidates. However, important results can be obtained by the use of other orthogonal functions such as Chebyshev polynomials.

5.3 Offset handling

ensuring the conditioning problem for the covariance matrix is not exacerbated; secondly they are easy to calculate (see for example [Rice 1964]).

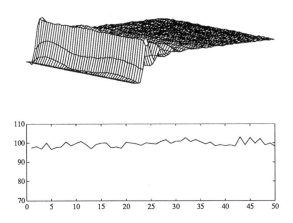

Fig 5.3.4 *Output with setpoint 100 for the static model, parameters known.*

In this case the reduction of order is justified and the self-tuning algorithm appears to converge to the optimal. Fig 5.3.7 shows the output when the order of ϕ is reduced further to include only the first four Legendre polynomials. It is clear that optimal control is not achieved. Fig 5.3.8 shows the true offset value and its estimated value for each of these two cases. Note that in this case we have not used integral control so the mean of each *column* of data is not necessarily that of the reference value. The use of integral control should rectify this, but because the model would remain underparametrised the optimal closed-loop behaviour would not be achieved.

From this discussion it should be clear that the solution for the incremental model is more satisfactory and more generally applicable. If the incremental model is inappropriate and we must use the static model then it seems likely that the problem of offset estimation is best tackled by considering the individual behaviour and constraints of the process itself.

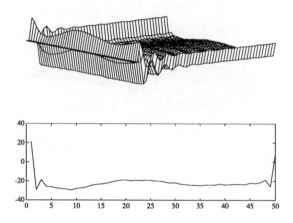

Fig 5.3.5 *Input with setpoint 100 for the static model, parameters known.*

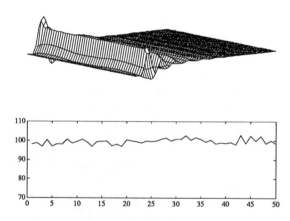

Fig 5.3.6 *Output with setpoint 100 for the static model, parameters estimated.*

5.3 Offset handling

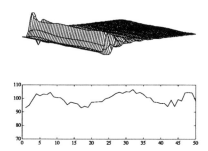

Fig 5.3.7 *Output with setpoint 100 for the static model, parameters estimated for reduced-order model.*

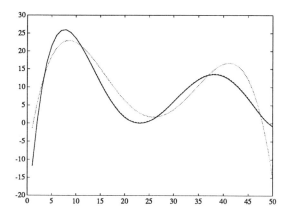

Fig 5.3.8 *Offset values. Solid line: true value. Dashed line: estimated value using fully parametrised model. Dotted line: estimated value using underparametrised model. Note that the dashed and solid lines coincide.*

Conclusion

In this work we have derived a generic class of self-tuning control algorithms for two-dimensional processes based on the one-dimensional generalised minimum variance controller [Clarke and Gawthrop 1975]. Our purpose was to develop two-dimensional controllers suitable for practical application. In §1.1 we identified two requirements for developing an appropriate theory. These were:

(1) The ability to model and parametrise the underlying dynamical process in an appropriate manner.

(2) The ability to design *simple* control strategies for the underlying process *based upon the model*.

To achieve these two goals we borrowed ideas from the image processing work of [Wagner 1987]. Specifically:

(1) We modelled the two-dimensional process as CARMA with non-symmetric half-plane causality. This allowed us to implement recursive least squares based parameter estimators while scanning through the data in a raster manner.

(2) We implemented controllers based on the partitioning of a two-dimensional CARMA process (Equation 2.4.3). This partition is a particular form of two-dimensional Diophantine equation with the key feature that it is simple to solve. Hence we avoided much of the notoriously difficult algebra associated with two-dimensional systems.

The solution of the partition (Equation 2.4.3) as it stands involves polynomials of infinite order. In [Wagner 1987] the prediction problem is solved by truncating the solution polynomials. In §2 we showed that the solution can be expressed in terms of finite-order *rational* polynomials and hence the least squares predictor can be implemented in an asymptotically optimal manner. This allowed us to develop minimum variance and

Conclusion

generalised minimum variance controllers which share the nice properties of the predictors of [Wagner 1987] and yet may be expressed in terms of 'classical' two-dimensional systems theory.

In §3 we turned our attention to the behaviour near the edges. This is rarely considered in the image processing theory since it is often reasonable to assume grey values away from the area of interest. For control, on the other hand, it is important to achieve good behaviour everywhere. Yet the 'classical' two-dimensional theory again tends to ignore the issue—this is because, strictly speaking, a process with such edges is no longer two-dimensional in the accepted sense. However we showed in §3.1 that while our algorithms require modifications to take into account the edge behaviour, the results of §2 in general remain valid when away from the edges. Furthermore, modelling such a process as multivariable in §3.3 allowed us to draw parallels between multivariable and two-dimensional theory.

In §4 we considered parameter estimation for two-dimensional processes in a general context. The algorithms of [Caldas-Pinto 1983] and [Wagner 1987] are a natural extension of the one-dimensional RELS or AML algorithms. However, the forgetting strategies employed are essentially one-dimensional. We attempted to resolve this limitation by developing a more general forgetting strategy, which included as a special case (row forgetting) an algorithm similar in behaviour to those of [Caldas-Pinto 1983] and [Wagner 1987]. In the forgetting strategy's most general form it is not possible to update the information matrix directly as is usual in one-dimensional recursive parameter schemes. Instead algorithms were developed based on updating the information matrix using QR factorisations. In simulations the general forgetting strategy was shown to be more versatile than that in [Wagner 1987], and also more versatile than the two special cases considered—namely row and column forgetting. However with large-order models it becomes hard to achieve a reasonable balance between quick response and good convergence properties. Thus where a priori knowledge is available about the type of change likely to occur the special cases of row and column forgetting become more powerful tools.

In §5 we showed that it is possible to combine the control algorithms of §3 with the parameter estimation schemes of §4 to synthesise self-tuning controllers. The asymptotic behaviour in simulations showed the usual nice properties associated with one-dimensional self-tuning controllers. We also considered some of the problems likely to be encountered in practical applications—specifically setpoint tracking and offset handling.

In both cases the simplest solution is to use integral control. However the algebraic properties of two-dimensional polynomials mean that this is not always a feasible solution. In such a case more computationally intensive solutions were offered.

Although there has been no attempt to apply the algorithms to real systems in the course of this work, the results of §5 encourage us in the belief that the algorithms and their associated theory are in a good state to begin such considerations. Indeed many of the examples were geared towards such an application to papermaking; in particular:

- The consideration of changing models near the edges in §3.3.7, §4.5.3 and §5.1.3.

- The modification of the controller for the case where the $B(w^{-1}, z^{-1})$ polynomial is no longer NSHP-causal (§3.4).

- The nature of the offset considered in §5.3.

These examples serve to show how the 'classical' theory must be modified to fit real situations, and yet how the practical problems may still be viewed from within a two-dimensional framework.

From a theoretical point of view the following issues remain outstanding:

- As in one dimension, the closed-loop stability of the generalised minimum variance self-tuning controller cannot be guaranteed. The use of a priori knowledge about likely models is essential if suitable weighting polynomials $P(w^{-1}, z^{-1})$ and $Q(w^{-1}, z^{-1})$ are to be chosen correctly. This problem is exacerbated in two dimensions where certain models (where the $A(w^{-1}, z^{-1})$ and $B(w^{-1}, z^{-1})$ polynomials share inverse unstable common zeros) cannot be stabilised.

- No consideration has been paid to the convergence properties of the self-tuning algorithms beyond observation of results in simulations. Convergence theory remains a thorny issue even in one dimension. In many of the simulations of §5 we combined a 2D-AML estimation algorithm with an explicit control algorithm resulting in a a closed-loop process whose input was not sufficiently exciting to ensure full rank of the covariance matrix. While it is possible to explain convergence in such a case heuristically (§5.1.1), without further analysis such a controller should only be applied in practice with extreme caution.

To conclude, we have developed a class of two-dimensional controllers which can be implemented in self-tuning form. While many theoretical issues remain outstanding their

Conclusion 173

simple nature together with good behaviour in simulations suggest that they may pave the way to the adoption of two-dimensional systems ideas in real control problems. It is by the success or failure of such future applications that they should be judged.

Appendix 1 Proof of the theorems

Theorem 2.2.1 *Let $y(m,n)$ be a stable ARMA process given by*

$$A(w^{-1}, z^{-1})y(m,n) = C(w^{-1}, z^{-1})e(m,n)$$

where

$$A(w^{-1}, z^{-1}) = \sum_{i=0}^{M}\sum_{j=0}^{N} a_{i,j} w^{-i} z^{-j} + \sum_{i=1}^{M}\sum_{j=1}^{N} a_{-i,j} w^{i} z^{-j}$$

and

$$C(w^{-1}, z^{-1}) = \sum_{i=0}^{M}\sum_{j=0}^{N} c_{i,j} w^{-i} z^{-j} + \sum_{i=1}^{M}\sum_{j=1}^{N} c_{-i,j} w^{i} z^{-j}$$

Define also $y'(m,n)$ such that

$$A'(w^{-1}, z^{-1})y'(m,n) = C'(w^{-1}, z^{-1})e'(m,n)$$

where

$$A'(w^{-1}, z^{-1}) = \sum_i \sum_j a'_{i,j} w^{-i} z^{-j}$$

$$C'(w^{-1}, z^{-1}) = \sum_i \sum_j c'_{i,j} w^{-i} z^{-j}$$

with

$$a'_{k_1 i + k_2 j, k_3 i + k_4 j} = a_{i,j}$$

$$c'_{k_1 i + k_2 j, k_3 i + k_4 j} = c_{i,j}$$

and with k_1, k_2, k_3 and k_4 all integers satisfying

$$k_1 k_4 - k_2 k_3 = \pm 1$$

Appendix 1 Proof of the theorems

Define also

$$r(i,j) = Ey(m,n)y(m-i,n-j)$$

$$r'(i,j) = Ey'(m,n)y'(m-i,n-j)$$

Then

$$r'(k_1 i + k_2 j, k_3 + k_4 j) = r(i,j)$$

Proof *From [O'Connor and Huang 1981] we have three important results:*

(1) y' is NSHP-causal and there is a $1-1$ onto mapping $A \to A'$, $C \to C'$.

(2) y stable $\Rightarrow$ y' stable.

(3) If

$$B = \frac{1}{A} = \sum_i \sum_j b_{i,j} w^{-i} z^{-j}$$

and

$$B' = \frac{1}{A'} = \sum_i \sum_j b'_{i,j} w^{-i} z^{-j}$$

then

$$b'_{k_1 i + k_2 j, k_3 i + k_4 j} = b_{i,j}$$

Given these three results, if we let

$$H = \frac{C}{A} = \sum_i \sum_j h_{i,j}$$

and

$$H' = \frac{C'}{A'} = \sum_i \sum_j h'_{i,j}$$

then

$$h'_{k_1 i + k_2 j, k_3 i + k_4 j} = \sum_{m'} \sum_{n'} b'_{k_1 i + k_2 j - m', k_3 i + k_4 j - n'} a'_{m', n'}$$

Since
$$k_1 k_4 - k_2 k_3 = \pm 1$$
we may uniquely define m and n by the relations
$$m' = k_1 m + k_2 n \text{ and } n' = k_3 m + k_4 n$$
so that
$$\begin{aligned}
h'_{k_1 i + k_2 j, k_3 i + k_4 j} &= \sum_m \sum_n b'_{k_1(i-m)+k_2(j-n), k_3(i-m)+k_4(j-n)} a'_{k_1 m + k_2 n, k_3 m + k_4 n} \\
&= \sum_m \sum_n b_{i-m, j-n} a_{m,n} \\
&= h_{i,j}
\end{aligned}$$

So
$$r'(k_1 i + k_2 j, k_3 i + k_4 j)$$

$$\begin{aligned}
&= E y'(m,n) y'(m - k_1 i - k_2 j, n - k_3 i - k_4 j) \\
&= \sum_{m'} \sum_{n'} h'_{m,n} h'_{m' - k_1 i - k_2 j, n' - k_3 i - k_4 j} \\
&= \sum_m \sum_n h'_{k_1 m + k_2 n, k_3 m + k_4 n} h'_{k_1(m-i)+k_2(n-j), k_3(m-i)+k_4(n-j)} \\
&= \sum_m \sum_n h_{m,n} h_{m-i, n-j} \\
&= r(i,j)
\end{aligned}$$

Theorem 2.4.1

There exist finite-order causal polynomials $\Phi^{k_1, k_2}(w^{-1}, z^{-1})$ and $\Gamma^{k_1, k_2}(w^{-1}, z^{-1})$ such that
$$F^{k_1, k_2}(w^{-1}, z^{-1}) = \frac{\Phi^{k_1, k_2}(w^{-1}, z^{-1})}{A(w^{-1}, 0)^{k_2}}$$

Appendix 1 Proof of the theorems 177

and

$$G^{k_1,k_2}(w^{-1}, z^{-1}) = \frac{\Gamma^{k_1,k_2}(w^{-1}, z^{-1})}{A(w^{-1}, 0)^{k_2}}$$

Proof We can express $A(w^{-1}, z^{-1})$ as

$$A(w^{-1}, z^{-1}) = A(w^{-1}, 0) \left[1 + \frac{1}{A(w^{-1}, 0)} \sum_{i=-M}^{M} \sum_{j=1}^{N} a_{i,j} w^{-i} z^{-j} \right]$$

so that

$$H(w^{-1}, z^{-1}) = \frac{C(w^{-1}, z^{-1})}{A(w^{-1}, z^{-1})}$$

$$= \frac{C(w^{-1}, z^{-1})}{A(w^{-1}, 0)} \sum_{p=0}^{\infty} \left[\frac{(-1)}{A(w^{-1}, 0)} \sum_{i=-M}^{M} \sum_{j=1}^{N} a_{i,j} w^{-i} z^{-j} \right]^p$$

Let

$$C(w^{-1}, z^{-1}) \left[\sum_{i=-M}^{M} \sum_{j=1}^{N} a_{i,j} w^{-i} z^{-j} \right]^p = \sum_{i=-(p+1)M}^{(p+1)M} \sum_{j=p}^{(p+1)N} s_{p,i,j} w^{-i} z^{-j}$$

for some $s_{p,i,j}$. Then for any $q \geq 0$

$$\sum_{i=-qM}^{\infty} h_{i,q} w^{-i} z^{-q} = \frac{1}{A(w^{-1}, 0)} \sum_{p=0}^{q} \left[\frac{(-1)}{A(w^{-1}, 0)} \right]^p \sum_{i=-(p+1)M}^{(p+1)M} s_{p,i,q} w^{-i} z^{-q}$$

Thus

$$F^{k_1,k_2}(w^{-1}, z^{-1}) = \sum_{q=0}^{k_2-1} \frac{1}{A(w^{-1}, 0)} \sum_{p=0}^{q} \left[\frac{(-1)}{A(w^{-1}, 0)} \right]^p \sum_{i=-(p+1)M}^{(p+1)M} s_{p,i,q} w^{-i} z^{-q}$$

$$+ \sum_{i=-k_2 M}^{k_1-1} h_{i,k_2} w^{-i} z^{-k_2}$$

and

$$A(w^{-1}, 0)^{k_2} F^{k_1,k_2}(w^{-1}, z^{-1})$$

$$= \sum_{q=0}^{k_2-1} \sum_{p=0}^{q} (-1)^p A(w^{-1}, 0)^{(k_2-p-1)} \sum_{i=-(p+1)M}^{(p+1)M} s_{p,i,q} w^{-i} z^{-q}$$

$$+ A(w^{-1}, 0)^{k_2} \sum_{i=-k_2 M}^{k_1-1} h_{i,k_2} w^{-i} z^{-k_2}$$

which is both causal and has a finite number of terms.

It is now simple to prove a similar result for

$$A(w^{-1},0)^{k_2} G^{k_1,k_2}(w^{-1},z^{-1})$$

Since $G^{k_1,k_2}(w^{-1},z^{-1})$ is causal

$$A(w^{-1},0)^{k_2} G^{k_1,k_2}(w^{-1},z^{-1})$$

must be so also. Furthermore

$$\begin{aligned} A(w^{-1},0)^{k_2} G^{k_1,k_2}(w^{-1},z^{-1}) \\ = w^{k_1} z^{k_2} A(w^{-1},0)^{k_2} C(w^{-1},z^{-1}) \\ - w^{k_1} z^{k_2} A(w^{-1},z^{-1}) A(w^{-1},0)^{k_2} F^{k_1,k_2}(w^{-1},z^{-1}) \end{aligned}$$

and so has a finite number of terms.

Theorem 2.4.2 *The least squares predictor for the two-dimensional CARMA process of §2.1 is given recursively by*

$$C(w^{-1},z^{-1}) \left[A(w^{-1},0)\right]^{k_2} \hat{y}(m+k_1, n+k_2|m,n) = \\ B(w^{-1},z^{-1}) \Phi^{k_1,k_2}(w^{-1},z^{-1}) u(m+k_1, n+k_2-\nu) + \Gamma^{k_1,k_2}(w^{-1},z^{-1}) y(m,n)$$

Proof *By orthogonality*

$$E\left[y(m+k_1, n+k_2) - y^*(m+k_1, n+k_2|m,n)\right]^2 \geq \\ E\left[F^{k_1,k_2}(w^{-1},z^{-1}) e(m+k_1, n+k_2)\right]^2$$

where $y^*(m+k_1, n+k_2|m,n)$ is any predictor given data up to (m,n) for the the future output $y(m+k_1, n+k_2)$. But

$$y(m+k_1, n+k_2) - \hat{y}(m+k_1, n+k_2|m,n)$$

$$= \frac{B}{A} u(m+k_1, n+k_2-\nu) + F^{k_1,k_2} e(m+k_1, n+k_2) + \frac{G^{k_1,k_2}}{A} e(m,n) \\ - \frac{B\Phi^{k_1,k_2}}{C\left[A(w^{-1},0)\right]^{k_2}} u(m+k_1, n+k_2-\nu) - \frac{\Gamma^{k_1,k_2}}{C\left[A(w^{-1},0)\right]^{k_2}} y(m,n)$$

Appendix 1 Proof of the theorems

$$\begin{aligned}
&= \left[\frac{B}{A} - w^{-k_1}z^{-k_2}\frac{BG^{k_1,k_2}}{AC} - \frac{BF^{k_1,k_2}}{C}\right]u(m+k_1, n+k_2-\nu) \\
&\quad + F^{k_1,k_2}e(m+k_1, n+k_2) + \left[\frac{G^{k_1,k_2}}{C} - \frac{G^{k_1,k_2}}{C}\right]y(m,n) \\
&= F^{k_1,k_2}e(m+k_1, n+k_2)
\end{aligned}$$

Hence $\hat{y}(m+k_1, n+k_2|m,n)$ is the optimal least squares predictor.

Theorem 2.4.3 *The variance of the prediction error V_{k_1,k_2} may be expressed as*

$$V_{k_1,k_2} = \sum_{\lambda=0}^{k_2-1} \text{var } y_\lambda(t) + \sum_{i=-\nu M}^{k_1-1} h_{i,k_2}^2$$

where

$$\text{var } y_\lambda(t) = \frac{1}{2\pi j}\oint_{|w|=1}\left[\frac{1}{[A(w^{-1},0)A(w,0)]^{k_2}}\frac{1}{w}\sum_{i=-\lambda M}^{k_2 M}\phi_{i,\lambda}w^{-i}\sum_{i=-\lambda M}^{k_2 M}\phi_{i,\lambda}w^i\right]dw$$

Proof *We know from §2.2.4 that*

$$V_{k_1,k_2} = \sigma^2 \sum_{j=0}^{k_2-1}\sum_{i=-jM}^{\infty} h_{i,j}^2 + \sigma^2 \sum_{i=-\nu M}^{k_1-1} h_{i,\nu}^2$$

But

$$\sum_{i=-jM}^{\infty} h_{i,j}w^{-i} = \frac{1}{[A(w^{-1},0)]^{k_2}}\sum_{i=-jM}^{k_2 M}\phi_{i,j}w^{-i}$$

where $\phi_{i,j}$ are the components of $\Phi^{k_1,k_2}(w^{-1}, z^{-1})$ for $0 \leq j \leq k_2 - 1$.

Consider the one-dimensional process given by

$$\begin{aligned}
y_j(t) &= \frac{1}{[A(w^{-1},0)]^{k_2}} \sum_{i=0}^{k_2 M + jM}\phi_{i-jM,j}w^{-i}e(t) \\
&= \frac{w^{-jM}}{[A(w^{-1},0)]^{k_2}} \sum_{i=-jM}^{k_2 M}\phi_{i,j}w^{-i}e(t) \\
&= w^{-jM} \sum_{i=-jM}^{\infty} h_{i,j}w^{-i}e(t)
\end{aligned}$$

So
$$\text{var } y_j(t) = \sigma^2 \sum_{i=-jM}^{\infty} h_{i,j}^2$$

Hence
$$V_{k_1,k_2} = \sum_{\lambda=0}^{k_2-1} \text{var } y_\lambda(t) + \sum_{i=-\nu M}^{k_1-1} h_{i,k_2}^2$$

where
$$\text{var } y_j(t) = \frac{1}{2\pi j} \oint_{|w|=1} \left[\frac{1}{[A(w^{-1},0)A(w,0)]^{k_2}} \frac{1}{w} \sum_{i=-\lambda M}^{k_2 M} \phi_{i,\lambda} w^{-i} \sum_{i=-\lambda M}^{k_2 M} \phi_{i,\lambda} w^{i} \right] dw$$

Theorem 2.5.1 *The control law for the criterion*
$$\min_{u(m,n)} Ey(m, n+\nu)^2$$
is to set
$$B(w^{-1}, z^{-1})\Phi^{0,\nu}(w^{-1}, z^{-1})u(m,n) + \Gamma^{0,\nu}(w^{-1}, z^{-1})y(m,n) = 0$$

Proof *Again by orthogonality we know*
$$Ey(m, n+\nu)^2 \geq E\left[F^{0,\nu}(w^{-1}, z^{-1})e(m, n+\nu)\right]^2$$
given any admissible input. We have equality when we set
$$B(w^{-1}, z^{-1})\Phi^{0,\nu}(w^{-1}, z^{-1})u(m,n) + \Gamma^{0,\nu}(w^{-1}, z^{-1})y(m,n) = 0$$
Hence this is the minimum variance controller.

Theorem 2.5.2 *The control law for the criterion*
$$\min_{u(m,n)} E_{(m,n)} \left\{ [Py(m, n+\nu) - Rr(m,n)]^2 + [Q'u(m,n)]^2 \right\}$$
is equivalent to the minimum variance control law
$$\min_{u(m,n)} E\phi(m, n+\nu)^2$$

Appendix 1 Proof of the theorems

where $\phi(m,n)$ is the pseudo-output given by

$$\phi(m,n) = Py(m,n) + z^{-\nu}Qu(m,n) - z^{-\nu}Rr(m,n)$$

and Q is given by

$$Q(w^{-1}, z^{-1}) = \frac{Q'(0,0)}{B(0,0)} Q'(w^{-1}, z^{-1})$$

Proof *Let*

$$J(m,n) = E_{(m,n)} \left\{ [Py(m, n+\nu) - Rr(m,n)]^2 + [Q'u(m,n)]^2 \right\}$$

so that the control criterion is

$$\min_{u(m,n)} J(m,n)$$

We may write

$$J = E_{(m,n)}\left\{[P\hat{y}(m, n+\nu|m,n) + PF^{0,\nu}e(m, n+\nu) - Rr(m,n)]^2 + [Q'u(m,n)]^2\right\}$$

Note that implicit in this notation is the identity

$$w^{-i}z^{-j}\hat{y}(m, n+\nu|m,n) = \hat{y}(m-i, n+\nu-j|m,n)$$

Similarly in the term $PF^{0,\nu}e(m, n+\nu)$ the polynomial $P(w^{-1}, z^{-1})$ should be viewed as an operator acting on $F^{0,\nu}e(m, n+\nu)$ with the identity

$$w^{-i}z^{-j}F^{0,\nu}e(m, n+\nu) = F^{-i,\nu-j}e(m-i, n+\nu-j)$$

Note also that

$$\hat{y}(m-i, n-j|m,n) = y(m-i, n-j) \text{ for } (m-i, n-j) \preceq (m,n)$$

and similarly

$$F^{-i,-j}e(m-i, n-j) = 0 \text{ for } (m-i, n-j) \preceq (m,n)$$

It follows that

$$E_{(m,n)}\left[P(w^{-1}, z^{-1})F^{0,\nu}e(m, n+\nu)\right]\left[P(w^{-1}, z^{-1})\hat{y}(m, n+\nu|m,n)\right] = 0$$

and similarly

$$E_{(m,n)}\left[P(w^{-1},z^{-1})F^{0,\nu}e(m,n+\nu)\right]\left[R(w^{-1},z^{-1})r(m,n)\right] = 0$$

Hence we may write

$$J = E_{(m,n)}\left\{[P\hat{y}(m,n+\nu|m,n) - Rr(m,n)]^2 + [Q'u(m,n)]^2\right\}$$
$$+ E_{(m,n)}\left\{[PF^{0,\nu}e(m,n+\nu)]^2\right\}$$

So that

$$\frac{\partial J}{\partial u(m,n)} = 2[P\hat{y}(m,n+\nu|m,n) - Rr(m,n)]\frac{\partial}{\partial u(m,n)}[P\hat{y}(m,n+\nu|m,n)]$$
$$+ 2Q'(0,0)Q'(w^{-1},z^{-1})u(m,n)$$

To determine

$$\frac{\partial}{\partial u(m,n)}[P\hat{y}(m,n+\nu|m,n)]$$

consider the expression given by Equation 2.4.6 for the predictor. It follows immediately that

$$\frac{\partial}{\partial u(m,n)}\left[w^{-i}z^{-j}\hat{y}(m,n+\nu|m,n)\right] = 0$$

for all $(m-i, n-j) \leq (m,n)$. Hence

$$\frac{\partial}{\partial u(m,n)}[P\hat{y}(m,n+\nu|m,n)] = P(0,0)\frac{\partial}{\partial u(m,n)}[\hat{y}(m,n+\nu|m,n)]$$

$$= B(0,0)$$

Hence the control law is

$$B(0,0)[P\hat{y}(m,n+\nu|m,n) - Rr(m,n)] + Q'(0,0)Q'(w^{-1},z^{-1})u(m,n) = 0$$

But this is precisely the minimum variance control law for the pseudo-output

$$\phi(m,n) = Py(m,n) + z^{-\nu}Qu(m,n) - z^{-\nu}Rr(m,n)$$

where

$$Q(w^{-1},z^{-1}) = \frac{Q'(0,0)}{B(0,0)}Q'(w^{-1},z^{-1})$$

Appendix 1 Proof of the theorems

Theorem 3.1.1 *The control law for the criterion*

$$\min_{u(m,n)} Ey(m, n + \nu)^2$$

is to estimate

$$\begin{aligned}\eta(m,n) &= A_m(w^{-1}, z^{-1})y(m,n) + B_m(w^{-1}, z^{-1})u(m, n - \nu) \\ &\quad - C'_m(w^{-1}, z^{-1})\eta(m,n)\end{aligned}$$

and then to set

$$X_{m,0,\nu}(w^{-1}, z^{-1})y(m,n) + Y_{m,0,\nu}(w^{-1}, z^{-1})u(m,n) + Z_{m,0,\nu}(w^{-1}, z^{-1})\eta(m,n)$$
$$= 0$$

Proof *In this case where there are finite edges we know that given data up to (m, n) then for any admissible input*

$$E\left[y(m, n + \nu)\right]^2 \geq E\left[F_{m,0,\nu}(w^{-1}, z^{-1})e(m, n + \nu)\right]^2$$

When we set the control law we have closed-loop output

$$y(m, n + \nu) = F_{m,0,\nu}(w^{-1}, z^{-1})e(m, n + \nu) + Z_{m,0,\nu}\left(e(m,n) - \eta(m,n)\right)$$

So

$$\begin{aligned}E\left[y(m, n + \nu)\right]^2 &= E\left[F_{m,0,\nu}(w^{-1}, z^{-1})e(m, n + \nu)\right]^2 \\ &\quad + \left[Z_{m,0,\nu}\left(e(m,n) - \eta(m,n)\right)\right]^2\end{aligned}$$

But since $C(w^{-1}, z^{-1})$ should be inverse stable (see §3.3.6) we have

$$e(m,n) - \eta(m,n) \to 0 \text{ as } n \to \infty$$

Furthermore $Z_{m,0,\nu}(w^{-1}, z^{-1})$ is fixed with finite order so

$$\left[Z_{m,0,\nu}\left(e(m,n) - \eta(m,n)\right)\right]^2 \to 0 \text{ as } n \to \infty$$

and so our controller converges to the optimal.

Theorem 3.1.3 *The control law for the criterion*

$$\min_{u(m,n)} E_{(m,n)} \left\{ [P_m y(m, n+\nu) - R_m r(m,n)]^2 + [Q'_m u(m,n)]^2 \right\}$$

is equivalent to the minimum variance control law

$$\min_{u(m,n)} E\psi(m, n+\nu)^2$$

where $\psi(m,n)$ is the pseudo-output given by

$$\psi(m,n) = P_m y(m,n) + z^{-\nu} Q_m u(m,n) - z^{-\nu} R_m r(m,n)$$

and Q_m is given by

$$Q_m(w^{-1}, z^{-1}) = \frac{Q'_m(0,0)}{B_m(0,0)} Q'_m(w^{-1}, z^{-1})$$

Proof *Our proof is broadly the same as for Theorem 2.5.2. We may write*

$$\begin{aligned} J &= E_{(m,n)} \left\{ [P_m \hat{y}(m, n+\nu|m, n) + P_m F_{m,0,\nu} e(m, n+\nu) - R_m r(m,n)]^2 \right. \\ &\quad \left. + [Q'_m u(m,n)]^2 \right\} \end{aligned}$$

As for Theorem 2.5.2 we have the identity

$$w^{-i} z^{-j} \hat{y}(m, n+\nu|m, n) = \hat{y}(m-i, n+\nu-j|m, n)$$

Again in the term $P_m F_{m,0,\nu} e(m, n+\nu)$ the polynomial $P(w^{-1}, z^{-1})$ should be viewed as an operator acting on $F_{m,0,\nu} e(m, n+\nu)$ with the identity

$$w^{-i} z^{-j} F_{m,0,\nu}(w^{-1}, z^{-1}) e(m, n+\nu) = F_{m,-i,\nu-j}(w^{-1}, z^{-1}) e(m-i, n+\nu-j)$$

Again

$$\hat{y}(m-i, n-j|m, n) = y(m-i, n-j) \text{ for } (m-i, n-j) \preceq (m,n)$$

and

$$F_{m,-i,-j}(w^{-1}, z^{-1}) e(m-i, n-j) = 0 \text{ for } (m-i, n-j) \preceq (m,n)$$

It follows again that

$$E_{(m,n)} [P_m(w^{-1}, z^{-1}) F_{m,0,\nu} e(m, n+\nu)] [P_m(w^{-1}, z^{-1}) \hat{y}(m, n+\nu|m, n)] = 0$$

Appendix 1 Proof of the theorems

and similarly

$$E_{(m,n)}\left[P_m(w^{-1}, z^{-1})F_{m,0,\nu}e(m, n+\nu)\right]\left[R_m(w^{-1}, z^{-1})r(m, n)\right] = 0$$

Hence we may write

$$J = E_{(m,n)}\left\{[P_m\hat{y}(m, n+\nu|m, n) - R_m r(m, n)]^2 + [Q'_m u(m, n)]^2\right\}$$
$$+ E_{(m,n)}\left\{[P_m F_{m,0,\nu}e(m, n+\nu)]^2\right\}$$

So

$$\frac{\partial J}{\partial u(m, n)} =$$
$$2\left[P_m\hat{y}(m, n+\nu|m, n) - R_m r(m, n)\right]\frac{\partial}{\partial u(m, n)}\left[P_m\hat{y}(m, n+\nu|m, n)\right]$$
$$+ 2Q'_m(0, 0)Q'_m(w^{-1}, z^{-1})u(m, n)$$

As in Theorem 2.5.2

$$\frac{\partial}{\partial u(m, n)}\left[P_m\hat{y}(m, n+\nu|m, n)\right] = P_m(0, 0)\frac{\partial}{\partial u(m, n)}\left[\hat{y}(m, n+\nu|m, n)\right]$$

and so

$$\frac{\partial}{\partial u(m, n)}\left[P_m\hat{y}(m, n+\nu|m, n)\right] = Y_{m,0,\nu}(0, 0)$$

But again due to the delay term $z^{-\nu}$ in the process Equation 3.1.2 we have straightforwardly

$$Y_{m,0,\nu}(0, 0) = B_m(0, 0)$$

Hence the control law is

$$B_m(0, 0)\left[P_m\hat{y}(m, n+\nu|m, n) - R_m r(m, n)\right]$$
$$+ 2Q'_m(0, 0)Q'_m(w^{-1}, z^{-1})u(m, n) = 0$$

But this is precisely the minimum variance control law for the pseudo-output

$$\psi(m, n) = P_m y(m, n) + z^{-\nu}Q_m u(m, n) - z^{-\nu}R_m r(m, n)$$

where

$$Q_m(w^{-1}, z^{-1}) = \frac{Q'_m(0, 0)}{B_m(0, 0)}Q'_m(w^{-1}, z^{-1})$$

Appendix 1 Proof of the theorems

Theorem 3.3.1 *The controller which minimises*

$$\min_{u(m,n)} Ey(m, n+\nu)^2$$

also minimises

$$\min_{u(1,n),\ldots,u(W,n)} E\sum_{i=1}^{W} y(i, n+\nu)^2$$

when applied successively.

Proof (by induction) *Let P(k) be the statement*

$$\min_{u(1,n)\ldots u(W,n)} E\sum_{i=1}^{W} y(i, n+\nu)^2 \geq$$

$$\min_{u(1,n)\ldots u(W-k,n)} E\sum_{i=1}^{W-k} y(i, n+\nu)^2 + \sum_{i=W-k+1}^{W} \min_{u(i,n)} Ey(i, n+\nu)^2$$

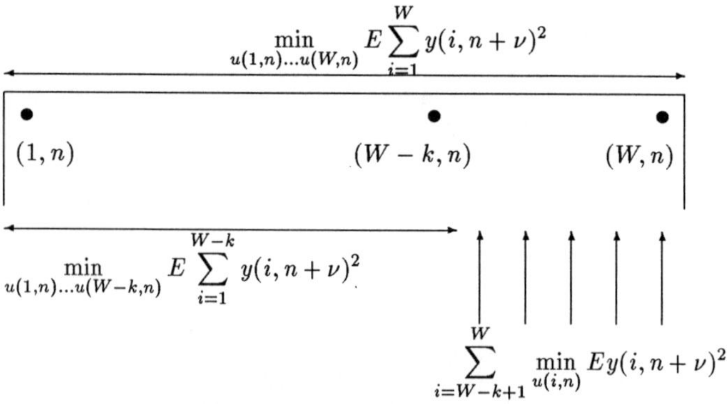

Fig A1.1 *Terms in P(k).*

Assume $P(k)$ is true for some k. We can break the second summation into two parts as:

$$\min_{u(1,n)\ldots u(W-k,n)} E\sum_{i=1}^{W-k} y(i, n+\nu)^2 =$$

$$\min_{u(1,n)\ldots u(W-k,n)} E\sum_{i=1}^{W-k-1} y(i, n+\nu)^2 + \min_{u(1,n)\ldots u(W-k,n)} Ey(W-k, n+\nu)^2$$

Appendix 1 Proof of the theorems

$y(i, n + \nu)$ for $1 \le i \le W - k - 1$

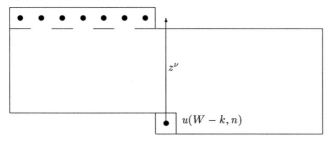

Fig A1.2 *The delay in the process.*

But due to the delay in the system the first part is independent of $u(W - k, n)$.
So

$$\min_{u(1,n)...u(W-k,n)} E \sum_{i=1}^{W-k-1} y(i, n+\nu)^2 = \min_{u(1,n)...u(W-k-1,n)} E \sum_{i=1}^{W-k-1} y(i, n+\nu)^2$$

Similarly for the second part

$$\min_{u(1,n)...u(W-k,n)} Ey(W-k, n+\nu)^2 \geq E\left[F_{W-k,0,\nu}(w^{-1}, z^{-1})e(W-k, n+\nu)\right]^2$$

$$= \min_{u(W-k,n)} Ey(W-k, n+\nu)^2$$

with equality achievable independently of $u(1,n)...u(W-k-1,n)$. *Hence*

$$\min_{u(1,n)...u(W-k,n)} E \sum_{i=1}^{W-k} y(i, n+\nu)^2 =$$

$$\min_{u(1,n)...u(W-k-1,n)} E \sum_{i=1}^{W-k-1} y(i, n+\nu)^2 + \min_{u(W-k,n)} Ey(W-k, n+\nu)^2$$

Thus we may restate $P(k)$:

$$\min_{u(1,n)...u(W,n)} E \sum_{i=1}^{W} y(i, n+\nu)^2$$

$$\geq \min_{u(1,n)...u(W-k,n)} E \sum_{i=1}^{W-k} y(i, n+\nu)^2 + \sum_{i=W-k+1}^{W} \min_{u(i,n)} Ey(i, n+\nu)^2$$

$$\geq \min_{u(1,n)...u(W-k-1,n)} E \sum_{i=1}^{W-k-1} y(i,n+\nu)^2 + \min_{u(W-k,n)} Ey(W-k,n+\nu)^2$$
$$+ \sum_{i=W-k+1}^{W} \min_{u(i,n)} Ey(i,n+\nu)^2$$
$$\geq \min_{u(1,n)...u(W-k-1,n)} E \sum_{i=1}^{W-k-1} y(i,n+\nu)^2 + \sum_{i=W-k}^{W} \min_{u(i,n)} Ey(i,n+\nu)^2$$
$$\geq \min_{u(1,n)...u(W-(k+1),n)} E \sum_{i=1}^{W-(k+1)} y(i,n+\nu)^2 + \sum_{i=W-(k+1)+1}^{W} \min_{u(i,n)} Ey(i,n+\nu)^2$$

Hence

$$P(k) \Rightarrow P(k+1)$$

$P(1)$ follows trivially and so by induction $P(k)$ is true for all k in the range $1 \leq k \leq W$; in particular $P(W)$ is true. So

$$\min_{u(1,n)...u(W,n)} E \sum_{i=1}^{W} y(i,n+\nu)^2 \geq \sum_{i=1}^{W} \min_{u(i,n)} Ey(i,n+\nu)^2$$

and we have equality when the controller for the criterion

$$\min_{u(m,n)} Ey(m,n+\nu)^2$$

is applied successively.

Theorem 3.3.2 *Suppose* $A(w^{-1},0) = 1$. *Then for W sufficiently large, and for both $m + k_1$ and $W - m - k_1$ sufficiently large*

$$C(w^{-1},z^{-1}) \stackrel{m+k_1}{\sim} \underline{C}(z^{-1})$$

$$F^{k_1,k_2}(w^{-1},z^{-1})B(w^{-1},z^{-1}) \stackrel{m+k_1}{\sim} \underline{C}(z^{-1})\underline{F}^m_{-k_2}(z^{-1})\left[\underline{C}(z^{-1})\right]^{-1}\underline{B}(z^{-1})$$

and

$$C(w^{-1},z^{-1}) - F^{k_1,k_2}(w^{-1},z^{-1})A(w^{-1},z^{-1}) \stackrel{m+k_1}{\sim}$$
$$\underline{C}(z^{-1}) - \underline{C}(z^{-1})\underline{F}^m_{-k_2}(z^{-1})\left[\underline{C}(z^{-1})\right]^{-1}\underline{A}(z^{-1})$$

Appendix 1 Proof of the theorems

Proof *From the construction of the polynomial matrices $\underline{\underline{A}}$, $\underline{\underline{B}}$ and $\underline{\underline{C}}$ (§3.1) we can label their respective (i,j)'th elements as $a'_{j-i}(z^{-1})$, $b'_{j-i}(z^{-1})$ and $c'_{j-i}(z^{-1})$. The corresponding two-dimensional polynomials for the case with edges are then*

$$A_m(w^{-1}, z^{-1}) = \sum_{i=\max(-M, m-W)}^{\min(M, m-1)} a'_i(z^{-1}) w^{-i}$$

and similarly for $B_m(w^{-1}, z^{-1})$ and $C_m(w^{-1}, z^{-1})$. It follows immediately that for $j-i > M$ and for $j-i < -M$

$$a'_{j-i}(z^{-1}) = b'_{j-i}(z^{-1}) = c'_{j-i}(z^{-1}) = 0$$

Furthermore we know from Equation 3.1.3 that for m sufficiently large and $W - m$ sufficiently large we have

$$\begin{aligned} A_m(w^{-1}, z^{-1}) &= \sum_{i=-M}^{M} a'_i(z^{-1}) w^{-i} \\ &= A(w^{-1}, z^{-1}) \end{aligned}$$

and similarly for $B_m(w^{-1}, z^{-1})$ and $C_m(w^{-1}, z^{-1})$. Hence the first part of our theorem: for both $m + k_1$ and $W - m - k_1$ large enough

$$C(w^{-1}, z^{-1}) \stackrel{m+k_1}{\sim} \underline{\underline{C}}(z^{-1})$$

Consider now the structure of $\underline{\underline{F}}_{k_2}^m$ with (i,j)'th element $f_{i,j}(z^{-1})$. We know by uniqueness (see Equation 3.3.4) that

$$F_{m,k_1,k_2}(w^{-1}, z^{-1}) = \sum_{i=1}^{W} f_{i, m+k_1}(z^{-1}) w^{i-m-k_1}$$

But we also know that $F_{m,k_1,k_2}(w^{-1}, z^{-1})$ has finite order so we may say there exists some finite θ such that for $j - i > \theta$ and for $j - i < -\theta$

$$f_{i,j}(z^{-1}) = 0$$

Furthermore for $m + k_1 - \theta \geq 1$ and $m + k_1 - \theta \leq W$ we know (Equation 3.1.8) that

$$F_{m,k_1,k_2}(w^{-1}, z^{-1}) = F^{k_1, k_2}(w^{-1}, z^{-1})$$

Hence we may define some $f'_i(z^{-1})$ such that for $j > \theta$ and $j \leq W - \theta$

$$f_{i,j}(z^{-1}) = f'_{j-i}(z^{-1})$$

In other words $\underline{\underline{F}}^m_{k_2}(z^{-1})$ has the structure

$$\underline{\underline{F}} = \begin{bmatrix} f_{1,1} & \cdots & f_{\theta+1,1} & & & & & 0 \\ & \ddots & & \ddots & & & & \\ f'_\theta & \cdots & f'_0 & \cdots & f'_{-\theta} & & & \\ & \ddots & & \ddots & & \ddots & & \\ & & & f'_\theta & \cdots & f'_0 & \cdots & f'_{-\theta} \\ & & & & \ddots & & \ddots & \\ 0 & & & & & f_{W-\theta,W} & \cdots & f_{W,W} \end{bmatrix}$$

We will also label the (i,j)'th elements of $\underline{\underline{C}}^{-1}$ and $\underline{\underline{C}}\,\underline{\underline{F}}^m_{k_2}\underline{\underline{C}}^{-1}\underline{\underline{B}}$ as $\gamma_{i,j}$ and $\beta_{i,j}$ respectively. Consider $\beta_{i,j}$:

$$\begin{aligned}
\beta_{i,j} &= \sum_{p=1}^{W}\sum_{q=1}^{W}\sum_{r=1}^{W} c'_{j-r} f_{q,r} \gamma_{p,q} b'_{p-i} \\
&= \sum_{p=1}^{W}\sum_{q=1}^{W}\sum_{r=-M}^{M} c'_r f_{q,j-r} \gamma_{p,q} b'_{p-i} \text{ provided } 1+M \leq j \leq W-M \\
&= \sum_{p=1}^{W}\sum_{q=1}^{W}\sum_{r=-M}^{M} c'_r f'_{j-r-q} \gamma_{p,q} b'_{p-i} \text{ provided } \begin{cases} 1+\theta \leq j-M \\ j+M \leq W-\theta \end{cases} \\
&= \sum_{p=1}^{W}\sum_{q=-\theta}^{\theta}\sum_{r=-M}^{M} c'_r f'_q \gamma_{p,j-r-q} b'_{p-i} \\
&= \sum_{p=1}^{W}\sum_{q=-\theta}^{\theta} \delta_{p,j-q} f'_q b'_{p-i} \text{ since } \underline{\underline{CC}}^{-1} = \underline{\underline{I}} \\
&= \sum_{q=-\theta}^{\theta} f'_q b'_{j-q-i}
\end{aligned}$$

But

$$F_{k_1,k_2}(w^{-1},z^{-1})B(w^{-1},z^{-1}) = \sum_{i=-\theta}^{\theta} f'_i(z^{-1}) w^{-i} \sum_{j=-M}^{M} b'_j(z^{-1}) w^{-j}$$

Appendix 1 Proof of the theorems

and so we have the second part of the theorem that

$$F^{k_1,k_2}(w^{-1},z^{-1})B(w^{-1},z^{-1}) \overset{m+k_1}{\sim} \underline{\underline{C}}(z^{-1})\underline{\underline{F}}^m_{k_2}(z^{-1})\left[\underline{\underline{C}}(z^{-1})\right]^{-1}\underline{\underline{B}}(z^{-1})$$

Furthermore since $\underline{\underline{A}}$ and $\underline{\underline{B}}$ have the same structure the third part follows immediately. That is

$$C(w^{-1},z^{-1}) - F^{k_1,k_2}(w^{-1},z^{-1})A(w^{-1},z^{-1}) \overset{m+k_1}{\sim}$$
$$\left[\underline{\underline{C}}(z^{-1}) - \underline{\underline{C}}(z^{-1})\underline{\underline{F}}^m_{k_2}(z^{-1})\left[\underline{\underline{C}}(z^{-1})\right]^{-1}\underline{\underline{A}}(z^{-1})\right]$$

Theorem 4.4.2 *Suppose*

$$\hat{\underline{\theta}} = \left[\lambda_1 \underline{\underline{X}}^T_1 \underline{\underline{X}}_1 + \lambda_2 \underline{\underline{X}}^T_2 \underline{\underline{X}}_2\right]^{-1} \left[\lambda_1 \underline{\underline{X}}^T_1 \underline{\underline{Y}}_1 + \lambda_2 \underline{\underline{X}}^T_2 \underline{\underline{Y}}_2\right]$$

where

$$\lambda_1, \lambda_2 \geq 0$$

and we already know

$$\begin{bmatrix} \underline{\underline{R}}_i & \underline{\underline{S}}_i \\ 0 & v_i \end{bmatrix} = \mathcal{R}\begin{bmatrix} \underline{\underline{X}}_i & \underline{Y}_i \end{bmatrix} \quad \text{for } i = 1,2$$

Then we have the two relations

$$\left[\lambda_1 \underline{\underline{X}}^T_1 \underline{\underline{X}}_1 + \lambda_2 \underline{\underline{X}}^T_2 \underline{\underline{X}}_2\right] = \underline{\underline{R}}^T \underline{\underline{R}}$$

and

$$\left[\lambda_1 \underline{\underline{X}}^T_1 \underline{Y}_1 + \lambda_2 \underline{\underline{X}}^T_2 \underline{Y}_2\right] = \underline{\underline{R}}^T \underline{S}$$

and hence

$$\hat{\underline{\theta}} = \underline{\underline{R}}^{-1} \underline{S}$$

where

$$\begin{bmatrix} \underline{\underline{R}} & \underline{S} \\ 0 & v \end{bmatrix} = \mathcal{R} \begin{bmatrix} \lambda_1^{\frac{1}{2}} \underline{\underline{R}}_1 & \lambda_1^{\frac{1}{2}} \underline{S}_1 \\ 0 & \lambda_1^{\frac{1}{2}} v_1 \\ \lambda_2^{\frac{1}{2}} \underline{\underline{R}}_2 & \lambda_2^{\frac{1}{2}} \underline{S}_2 \\ 0 & \lambda_2^{\frac{1}{2}} v_2 \end{bmatrix}$$

Proof *There exist some orthogonal $\underline{\underline{Q}}$ and $\underline{\underline{Q}}_i$ for $i = 1, 2$ such that*

$$\underline{\underline{R}} = \underline{\underline{Q}}^T \begin{bmatrix} \lambda_1^{\frac{1}{2}} \underline{\underline{R}}_1 \\ \lambda_2^{\frac{1}{2}} \underline{\underline{R}}_2 \end{bmatrix}$$

$$\underline{\underline{S}} = \underline{\underline{Q}}^T \begin{bmatrix} \lambda_1^{\frac{1}{2}} \underline{\underline{S}}_1 \\ \lambda_2^{\frac{1}{2}} \underline{\underline{S}}_2 \end{bmatrix}$$

$$\underline{\underline{R}}_i = \underline{\underline{Q}}_i^T \underline{\underline{X}}_i$$

and

$$\underline{\underline{S}}_i = \underline{\underline{Q}}_i^T \underline{\underline{Y}}$$

So

$$\begin{aligned}
\underline{\underline{R}}^T \underline{\underline{R}} &= \begin{bmatrix} \lambda_1^{\frac{1}{2}} \underline{\underline{R}}_1 \\ \lambda_2^{\frac{1}{2}} \underline{\underline{R}}_2 \end{bmatrix}^T \underline{\underline{Q}} \, \underline{\underline{Q}}^T \begin{bmatrix} \lambda_1^{\frac{1}{2}} \underline{\underline{R}}_1 \\ \lambda_2^{\frac{1}{2}} \underline{\underline{R}}_2 \end{bmatrix} \\
&= \begin{bmatrix} \lambda_1^{\frac{1}{2}} \underline{\underline{R}}_1 \\ \lambda_2^{\frac{1}{2}} \underline{\underline{R}}_2 \end{bmatrix}^T \begin{bmatrix} \lambda_1^{\frac{1}{2}} \underline{\underline{R}}_1 \\ \lambda_2^{\frac{1}{2}} \underline{\underline{R}}_2 \end{bmatrix} \\
&= \lambda_1 \underline{\underline{R}}_1^T \underline{\underline{R}}_1 + \lambda_2 \underline{\underline{R}}_2^T \underline{\underline{R}}_2 \\
&= \lambda_1 \underline{\underline{X}}_1^T \underline{\underline{Q}}_1 \underline{\underline{Q}}_1^T \underline{\underline{X}}_1 + \lambda_2 \underline{\underline{X}}_2^T \underline{\underline{Q}}_2 \underline{\underline{Q}}_2^T \underline{\underline{X}}_2 \\
&= \lambda_1 \underline{\underline{X}}_1^T \underline{\underline{X}}_1 + \lambda_2 \underline{\underline{X}}_2^T \underline{\underline{X}}_2
\end{aligned}$$

Similarly

$$\begin{aligned}
\underline{\underline{R}}^T \underline{\underline{S}} &= \begin{bmatrix} \lambda_1^{\frac{1}{2}} \underline{\underline{R}}_1 \\ \lambda_2^{\frac{1}{2}} \underline{\underline{R}}_2 \end{bmatrix}^T \underline{\underline{Q}} \, \underline{\underline{Q}}^T \begin{bmatrix} \lambda_1^{\frac{1}{2}} \underline{\underline{S}}_1 \\ \lambda_2^{\frac{1}{2}} \underline{\underline{S}}_2 \end{bmatrix} \\
&= \begin{bmatrix} \lambda_1^{\frac{1}{2}} \underline{\underline{R}}_1 \\ \lambda_2^{\frac{1}{2}} \underline{\underline{R}}_2 \end{bmatrix}^T \begin{bmatrix} \lambda_1^{\frac{1}{2}} \underline{\underline{S}}_1 \\ \lambda_2^{\frac{1}{2}} \underline{\underline{S}}_2 \end{bmatrix} \\
&= \lambda_1 \underline{\underline{R}}_1^T \underline{\underline{S}}_1 + \lambda_2 \underline{\underline{R}}_2^T \underline{\underline{S}}_2
\end{aligned}$$

$$= \lambda_1 \underline{X}_1^T \underline{\underline{Q}}_1 \underline{\underline{Q}}_1^T \underline{Y}_1 + \lambda_2 \underline{X}_2^T \underline{\underline{Q}}_2 \underline{\underline{Q}}_2^T \underline{Y}_2$$

$$= \lambda_1 \underline{X}_1^T \underline{Y}_1 + \lambda_2 \underline{X}_2^T \underline{Y}_2$$

It follows immediately that

$$\hat{\underline{\theta}} = \left[\lambda_1 \underline{X}_1^T \underline{X}_1 + \lambda_2 \underline{X}_2^T \underline{X}_2 \right]^{-1} \left[\lambda_1 \underline{X}_1^T \underline{Y}_1 + \lambda_2 \underline{X}_2^T \underline{Y}_2 \right]$$

$$= \left[\underline{\underline{R}}^T \underline{\underline{R}} \right]^{-1} \left[\underline{\underline{R}}^T \underline{S} \right]$$

$$= \underline{\underline{R}}^{-1} \underline{S}$$

Theorem 4.4.3 *Suppose*

$$\hat{\underline{\theta}} = \left[\lambda_1 \underline{X}_1^T \underline{X}_1 + \lambda_2 \underline{X}_2^T \underline{X}_2 \right]^{-1} \left[\lambda_1 \underline{X}_1^T \underline{Y}_1 + \lambda_2 \underline{X}_2^T \underline{Y}_2 \right]$$

and we already know

$$\left\{ \underline{D}_i, \begin{bmatrix} \underline{\underline{R}}_i & \underline{S}_i \\ 0 & v_i \end{bmatrix} \right\} = \mathcal{G} \left\{ \underline{I}, \begin{bmatrix} \underline{X}_i & \underline{Y}_i \end{bmatrix} \right\} \text{ for } i = 1, 2$$

Then we have the two relations

$$\left[\lambda_1 \underline{X}_1^T \underline{X}_1 + \lambda_2 \underline{X}_2^T \underline{X}_2 \right] = \underline{\underline{R}}^T \underline{\underline{D}} \underline{\underline{R}}$$

and

$$\left[\lambda_1 \underline{X}_1^T \underline{Y}_1 + \lambda_2 \underline{X}_2^T \underline{Y}_2 \right] = \underline{\underline{R}}^T \underline{\underline{D}} \underline{S}$$

and hence

$$\hat{\underline{\theta}} = \underline{\underline{R}}^{-1} \underline{S}$$

where

$$\left\{ \underline{D}, \begin{bmatrix} \underline{\underline{R}} & \underline{S} \\ 0 & v \end{bmatrix} \right\} = \mathcal{G} \left\{ \begin{bmatrix} \lambda_1 \underline{\underline{D}}_1 & 0 \\ 0 & \lambda_2 \underline{\underline{D}}_2 \end{bmatrix}, \begin{bmatrix} \underline{\underline{R}}_1 & \underline{S}_1 \\ 0 & v_1 \\ \underline{\underline{R}}_2 & \underline{S}_2 \\ 0 & v_2 \end{bmatrix} \right\}$$

Proof *There exist some orthogonal $\underline{\underline{Q}}$ and $\underline{\underline{Q}}_i$ for $i = 1, 2$ such that*

$$\underline{\underline{D}}^{\frac{1}{2}}\underline{\underline{R}} = \underline{\underline{Q}}^T \begin{bmatrix} \lambda_1^{\frac{1}{2}}\underline{\underline{D}}_1^{\frac{1}{2}} & 0 \\ 0 & \lambda_2^{\frac{1}{2}}\underline{\underline{D}}_2^{\frac{1}{2}} \end{bmatrix} \begin{bmatrix} \underline{\underline{R}}_1 \\ \underline{\underline{R}}_2 \end{bmatrix}$$

$$\underline{\underline{D}}^{\frac{1}{2}}\underline{\underline{S}} = \underline{\underline{Q}}^T \begin{bmatrix} \lambda_1^{\frac{1}{2}}\underline{\underline{D}}_1^{\frac{1}{2}} & 0 \\ 0 & \lambda_2^{\frac{1}{2}}\underline{\underline{D}}_2^{\frac{1}{2}} \end{bmatrix} \begin{bmatrix} \underline{S}_1 \\ \underline{S}_2 \end{bmatrix}$$

$$\underline{\underline{D}}_i^{\frac{1}{2}}\underline{\underline{R}}_i = \underline{\underline{Q}}_i^T \underline{\underline{X}}_i$$

and

$$\underline{\underline{D}}_i^{\frac{1}{2}}\underline{S}_i = \underline{\underline{Q}}_i^T \underline{Y}$$

So

$$\underline{\underline{R}}^T \underline{\underline{D}} \underline{\underline{R}} = \begin{bmatrix} \underline{\underline{R}}_1 \\ \underline{\underline{R}}_2 \end{bmatrix}^T \begin{bmatrix} \lambda_1\underline{\underline{D}}_1 & 0 \\ 0 & \lambda_2\underline{\underline{D}}_2 \end{bmatrix} \begin{bmatrix} \underline{\underline{R}}_1 \\ \underline{\underline{R}}_2 \end{bmatrix}$$

$$= \lambda_1 \underline{\underline{R}}_1^T \underline{\underline{D}}_1 \underline{\underline{R}}_1 + \lambda_2 \underline{\underline{R}}_2^T \underline{\underline{D}}_2 \underline{\underline{R}}_2$$

$$= \lambda_1 \underline{\underline{X}}_1^T \underline{\underline{X}}_1 + \lambda_2 \underline{\underline{X}}_2^T \underline{\underline{X}}_2$$

Similarly

$$\underline{\underline{R}}^T \underline{\underline{D}} \underline{\underline{S}} = \begin{bmatrix} \underline{\underline{R}}_1 \\ \underline{\underline{R}}_2 \end{bmatrix}^T \begin{bmatrix} \lambda_1\underline{\underline{D}}_1 & 0 \\ 0 & \lambda_2\underline{\underline{D}}_2 \end{bmatrix} \begin{bmatrix} \underline{S}_1 \\ \underline{S}_2 \end{bmatrix}$$

$$= \lambda_1 \underline{\underline{R}}_1^T \underline{\underline{D}}_1 \underline{S}_1 + \lambda_2 \underline{\underline{R}}_2^T \underline{\underline{D}}_2 \underline{S}_2$$

$$= \lambda_1 \underline{\underline{X}}_1^T \underline{Y}_1 + \lambda_2 \underline{\underline{X}}_2^T \underline{Y}_2$$

It follows immediately that

$$\underline{\hat{\theta}} = \left[\lambda_1 \underline{\underline{X}}_1^T \underline{\underline{X}}_1 + \lambda_2 \underline{\underline{X}}_2^T \underline{\underline{X}}_2 \right]^{-1} \left[\lambda_1 \underline{\underline{X}}_1^T \underline{Y}_1 + \lambda_2 \underline{\underline{X}}_2^T \underline{Y}_2 \right]$$

$$= \left[\underline{\underline{R}}^T \underline{\underline{D}} \underline{\underline{R}} \right]^{-1} \left[\underline{\underline{R}}^T \underline{\underline{D}} \underline{\underline{S}} \right]$$

$$= \underline{\underline{R}}^{-1} \underline{S}$$

Appendix 1 Proof of the theorems

Theorem 5.2.1 *If $B(w^{-1}, z^{-1})$ is symmetric then we can introduce integral control of the form $(1 - z^{-1})$ if and only if all factors of*

$$B|_{z^{-1}=1} = B_0 \prod_i (1 + \beta_i w^{-1} + \beta_i w)$$

have

$$|\beta| < 0.5$$

Proof Let

$$B(w^{-1}, z^{-1}) = \sum_{j=0}^{N} b_{0,j} z^{-j} + \sum_{i=1}^{M} \sum_{j=1}^{N} b_{i,j}(w^{-i} + w^i) z^{-j}$$

We are interested in zeros of B when $z^{-1} = 1$. First we must transform

$$w^M z^{-1} \to \zeta^{-1}$$

so that $B \to B'$ where B' is quarter-plane causal, while the stability characteristics of B are preserved (§2.2.1).

$$B'(w^{-1}, \zeta^{-1}) = \sum_{j=0}^{N} b_{0,j} w^{-jM} \zeta^{-j} + \sum_{i=1}^{M} \sum_{j=1}^{N} b_{i,j}(w^{-i-jM} + w^{i-jM}) \zeta^{-j}$$

We are now interested in zeros of B' when $w^{-M} \zeta^{-1} = 1$.

$$B'|_{w^{-M}\zeta^{-1}=1} = \sum_{j=0}^{N} b_{0,j} + \sum_{i=1}^{M} \sum_{j=1}^{N} b_{i,j}(w^{-i} + w^i)$$

Suppose w_0^{-1} is a zero; then so is w_0. Hence we may write

$$B'|_{w^{-M}\zeta^{-1}=1} = B_0 \prod_i (1 + \beta_i w^{-1} + \beta_i w)$$

for some B_0 and β_i.

Suppose $w_0^{-1} = re^{i\phi}$ is a zero of $1 + \beta w^{-1} + \beta w$. Then

$$1 + \beta r e^{i\phi} + \frac{\beta}{r} e^{-i\phi} = 0$$

So taking real and imaginary parts

$$1 + \beta(r + \frac{1}{r})\cos\phi = 0$$

and

$$\beta(r - \frac{1}{r})\sin\phi = 0$$

We may distinguish two cases. In the first case

$$r - \frac{1}{r} = 0$$

in which case

$$|w_0| = 1$$

so we have a marginally stable root. In the second case

$$\sin\phi = 0$$

Then either

$$1 + \beta(r + \frac{1}{r}) = 0$$

or

$$1 - \beta(r + \frac{1}{r}) = 0$$

So

$$|\beta| \leq 0.5 \text{ with equality if and only if } |r| = 1$$

We may summarise the two cases as the following:
If

$$|\beta| \geq 0.5$$

then we have a marginally stable common zero at

$$|w_0^{-1}| = 1 \text{ and } |w_0^{-M}z_0^{-1}| = 1$$

But if

$$|\beta| < 0.5$$

Appendix 1 Proof of the theorems

then for any common zero

$$|w_0^{-1}| \neq 1$$

In this second case if

$$|w_0^{-1}| > 1$$

then we have a stable common zero while if

$$|w_0^{-1}| < 1$$

then

$$|\zeta_0^{-1}| > 1$$

so again we have a common stable zero.

Hence we have the result that when $B(w^{-1}, z^{-1})$ is symmetric we can introduce integral control of the form $(1 - z^{-1})$ if and only if all factors of

$$B|_{z^{-1}=1} = B_0 \prod_i (1 + \beta_i w^{-1} + \beta_i w)$$

have

$$|\beta| < 0.5$$

Appendix 2 Algorithm 2.4.1

Here we present an algorithm for the calculation of the polynomial $\Phi^{k_1,k_2}(w^{-1},z^{-1})$ defined in Equation 2.4.4.

Combining Equations 2.4.3, 2.4.4 and 2.4.5 we have the identity

$$\left[A(w^{-1},0)\right]^{k_2} C(w^{-1},z^{-1}) = A(w^{-1},z^{-1})\Phi(w^{-1},z^{-1})$$
$$+ w^{-k_1} z^{-k_2} \Gamma(w^{-1},z^{-1}) \tag{A2.1}$$

The presence of z^{-k_2} on the righthand side of Equation A2.1 allows us to calculate $\phi_{i,j}$ for $j \leq k_2 - 1$ simply by equating coefficients. Firstly for $j = 0$ we have

$$A(w^{-1},0) \sum_i \phi_{i,0} w^{-i} = \left[A(w^{-1},0)\right]^{k_2} \sum_{i=0}^{M} c_{i,0} w^{-i}$$

and hence

$$\sum_{i=0}^{k_2 M} \phi_{i,0} w^{-i} = [A(w^{-1},0)]^{(k_2-1)} \sum_{i=0}^{M} c_{i,0} w^{-i} \tag{A2.2}$$

Then for $1 \leq j \leq k_2 - 1$ (assuming $k_2 \geq 2$) we have the relation

$$A(w^{-1},0) \sum_i \phi_{i,j} w^{-i} = \left[A(w^{-1},0)\right]^{k_2} \sum_{i=-M}^{M} c_{i,j} w^{-i}$$
$$- \sum_{\lambda=1}^{j-1} \left[\sum_{i=-M}^{M} a_{i,\lambda} w^{-i}\right] \left[\sum_i \phi_{i,j-\lambda} w^{-i}\right] - \left[\sum_{i=-M}^{M} a_{i,j} w^{-i}\right] \left[\sum_{i=0}^{\nu M} \phi_{i,0} w^{-i}\right]$$

If we now define $f^*_{i,j}$ such that

$$\sum_i \phi_{i,j} w^{-i} = [A(w^{-1},0)]^{(k_2-j-1)} \sum_i f^*_{i,j} w^{-i}$$

then

$$[A(w^{-1},0)]^{(k_2-j)} \sum_i f^*_{i,j} w^{-i}$$
$$= [A(w^{-1},0)]^{k_2} \sum_{i=-M}^{M} c_{i,j} w^{-i}$$

Appendix 2 Algorithm 2.4.1 199

$$-\sum_{\lambda=1}^{j-1}\left[\sum_{i=-M}^{M}a_{i,\lambda}w^{-i}\right][A(w^{-1},0)]^{(k_2+\lambda-j-1)}\left[\sum_{i}f^*_{i,j-\lambda}w^{-i}\right]$$

$$-\left[\sum_{i=-M}^{M}a_{i,j}w^{-i}\right][A(w^{-1},0)]^{(\nu-1)}\left[\sum_{i=0}^{M}c_{i,0}w^{-i}\right]$$

and hence

$$\sum_{i=-jM}^{(j+1)M} f^*_{i,j}w^{-i} = [A(w^{-1},0)]^j \sum_{i=-M}^{M} c_{i,j}w^{-i}$$

$$-\sum_{\lambda=1}^{j-1}\left[\sum_{i=-M}^{M}a_{i,\lambda}w^{-i}\right][A(w^{-1},0)]^{(\lambda-1)}\left[\sum_{i=(\lambda-j)M}^{(j-\lambda+1)M}f^*_{i,j-\lambda}w^{-i}\right]$$

$$-\left[\sum_{i=-M}^{M}a_{i,j}w^{-i}\right][A(w^{-1},0)]^{(j-1)}\left[\sum_{i=0}^{M}c_{i,0}w^{-i}\right] \quad\quad (A2.3)$$

We can thus find the $f^*_{i,j}$'s by equating coefficients and hence find the $\phi_{i,j}$'s. We have then only to calculate the ϕ_{i,k_2}'s. We can do this by first finding the h_{i,k_2} terms for $-k_2 M \leq i \leq k_1 - 1$ from the expansion of

$$\frac{C(w^{-1},z^{-1})}{A(w^{-1},z^{-1})} = \sum_{i,j} h_{i,j} w^{-i} z^{-j}$$

Hence our algorithm:

Algorithm 2.4.1. (Calculation of $\Phi^{k_1,k_2}(w^{-1},z^{-1})$)

(1) Calculate $\phi_{i,0}$ for $0 \leq i \leq k_2 M$ from Equation A2.2.

(2) For $1 \leq j \leq k_2 - 1$ (assuming $k_2 \geq 2$) calculate the $f^*_{i,j}$'s from Equation A2.3. The corresponding $\phi_{i,j}$'s can then be found using the relationship

$$\sum_{i=-jM}^{k_2 M} \phi_{i,j} w^{-i} = [A(w^{-1},0)]^{(k_2-j-1)} \sum_{i=-jM}^{(j+1)M} f^*_{i,j} w^{-i}$$

(3) Calculate the f^*_{i,k_2}'s according to Equation A2.3.

(4) Make the expansion

$$\sum_{i=-k_2 M}^{\infty} h_{i,k_2} w^{-i} = \frac{1}{[A(w^{-1},0)]^{(k_2+1)}} \sum_{i=-k_2 M}^{(k_2+1)M} f^*_{i,k_2} w^{-i}$$

for the necessary number of terms.

(5) Calculate

$$\sum_{i=-k_2 M}^{k_1-1+k_2 M} \phi_{i,k_2} w^{-i} = [A(w^{-1},0)]^{k_2} \sum_{i=-k_2 M}^{k_1-1} h_{i,k_2}$$

Appendix 3 Multivariable self-tuning control

Here we present the pertinent results from [Borisson 1979] and [Koivo 1980]. We will adapt their notation to fit with the main body of our text.

A3.1 The process

We will consider the multivariable process

$$\underline{A}(z^{-1})\underline{y}(n) = z^{-\nu}\underline{B}(z^{-1})\underline{u}(n) + \underline{C}(z^{-1})\underline{e}(n)$$

where $\underline{A}(z^{-1})$, $\underline{B}(z^{-1})$ and $\underline{C}(z^{-1})$ are W by W polynomial matrices and $\underline{y}(n)$, $\underline{u}(n)$ and $\underline{e}(n)$ are W term vectors. We require $\det \underline{C}(z)$ to have all its zeros outside the unit disc. For minimum variance control (§3.3) we will impose the same condition on $\det \underline{B}(z)$. We also require $\underline{B}(0)$ to be non-singular.

A3.2 Prediction

Here we give the least squares optimal predictor $\hat{\underline{y}}(n+k|n)$ where $k \leq \nu$. Partition

$$\underline{C}(z^{-1}) = \underline{A}(z^{-1})\underline{F}_k(z^{-1}) + z^{-k}\underline{G}_k \qquad (A3.1)$$

where the order of $\underline{F}_k(z^{-1})$ is $k-1$. There exist nonunique matrices $\underline{\tilde{F}}_k(z^{-1})$ and $\underline{\tilde{G}}_k(z^{-1})$ such that

$$\underline{\tilde{F}}_k(z^{-1})\underline{G}_k(z^{-1}) = \underline{\tilde{G}}_k(z^{-1})\underline{F}_k(z^{-1})$$

and

$$\det \underline{\tilde{F}}_k(z^{-1}) = \det \underline{F}_k(z^{-1})$$

and

$$\underline{\tilde{F}}_k(0) = \underline{I}$$

Define the polynomial matrix $\underline{\tilde{C}}_k(z^{-1})$ as

$$\underline{\tilde{C}}_k(z^{-1}) = \underline{\tilde{F}}_k(z^{-1})\underline{A}(z^{-1}) + z^{-k}\underline{\tilde{G}}_k(z^{-1})$$

so that

$$\underline{\tilde{C}}_k(z^{-1})\underline{F}_k(z^{-1}) = \underline{\tilde{F}}_k(z^{-1})\underline{C}(z^{-1})$$

Then the least squares optimal predictor can be shown to be

$$\underline{\hat{y}}(n+k|n) = \left[\underline{\tilde{C}}_k(z^{-1})\right]^{-1}\left[\underline{\tilde{G}}_k(z^{-1})\underline{y}(n) + \underline{\tilde{F}}_k(z^{-1})\underline{B}(z^{-1})\underline{u}(n)\right]$$

A3.3 Minimum variance control

As usual the minimum variance control law is obtained by setting the ν-step predictor to zero. Specifically the control law is to set

$$\underline{\tilde{G}}_\nu(z^{-1})\underline{y}(n) + \underline{\tilde{F}}_\nu(z^{-1})\underline{B}(z^{-1})\underline{u}(n) = 0$$

However we should be specific about what it is we are minimising. An admissible input is such that $\underline{u}(n)$ is a function of $\underline{y}(n-i)$ for $i \geq 0$ and $\underline{u}(n-i)$ for $i > 0$. Our control criterion is then to minimise over any admissible input the cost function

$$J = E\left[\underline{y}^T(n+\nu)\underline{Q}\,\underline{y}(n+\nu)\right]$$

where $\underline{Q}$ is any positive semidefinite matrix.

The closed-loop output is then

$$\underline{y}(n) = \underline{F}_\nu(z^{-1})\underline{e}(n)$$

with corresponding input

$$\underline{u}(n) = -\left[\underline{B}(z^{-1})\right]^{-1}\underline{G}_\nu(z^{-1})\underline{e}(n)$$

A3.4 Generalised minimum variance control

The control criterion for the generalised minimum variance controller is to minimise over any admissible input (as defined above) the cost function

$$J = E\left\{\left\|\underline{P}(z^{-1})\underline{y}(n+\nu) - \underline{R}(z^{-1})\underline{r}(n)\right\|^2 + \left\|\underline{Q}'(z^{-1})\underline{u}(n)\right\|^2\right\}$$

where $\underline{r}(n)$ is the setpoint. Notice that the weighting operator $P(z^{-1})$ is restricted to being a polynomial in z^{-1} (as opposed to a more general polynomial matrix). It can be shown that minimising this cost function is equivalent to the minimum variance control law for the pseudo-output $\underline{\phi}(n)$ given by

$$\underline{\phi}(n) = P(z^{-1})\underline{y}(n) - \underline{R}(z^{-1})\underline{r}(n-\nu) + \underline{Q}(z^{-1})\underline{u}(n-\nu)$$

where

$$\underline{\underline{Q}}(z^{-1}) = \left[\underline{\underline{B}}^T(0)\right]^{-1} \underline{\underline{Q}}'^T(0)\underline{\underline{Q}}'(z^{-1})$$

Appendix 4 Simulation results for §4.6

Here we present the remaining simulation results from §4.6. The Figures are all listed in Tables 4.6.2 and 4.6.3.

Fig A4.1 *'Mesh' of $\hat{a}$ for case (ii) using general two-dimensional forgetting.*

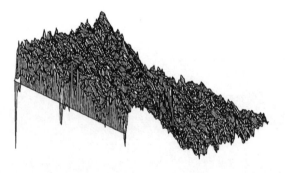

Fig A4.2 *'Mesh' of $\hat{c}$ for case (ii) using general two-dimensional forgetting.*

Appendix 4 Simulation results for §4.6

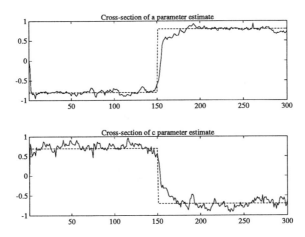

Fig A4.3 *Vertical cross-section of â and ĉ for case (ii) using general two-dimensional forgetting.*

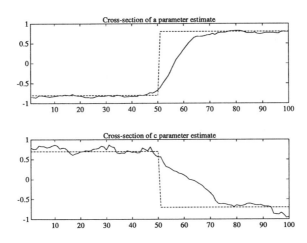

Fig A4.4 *Horizontal cross-section of â and ĉ for case (ii) using general two-dimensional forgetting.*

206 Appendix 4 Simulation results for §4.6

Fig A4.5 *'Mesh' of â for case (iv) using general two-dimensional forgetting.*

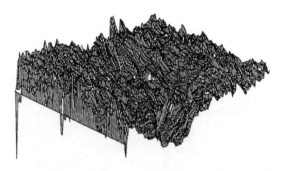

Fig A4.6 *'Mesh' of ĉ for case (iv) using general two-dimensional forgetting.*

Appendix 4 Simulation results for §4.6

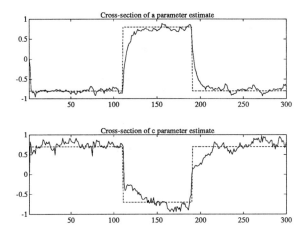

Fig A4.7 *Vertical cross-section of $\hat{a}$ and $\hat{c}$ for case (iv) using general two-dimensional forgetting.*

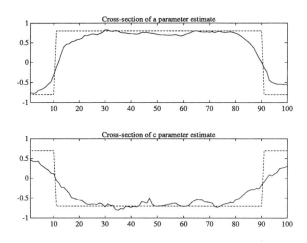

Fig A4.8 *Horizontal cross-section of $\hat{a}$ and $\hat{c}$ for case (iv) using general two-dimensional forgetting.*

Appendix 4 Simulation results for §4.6

Fig A4.9 *'Mesh' of â for case (ii) using Wagner's algorithm.*

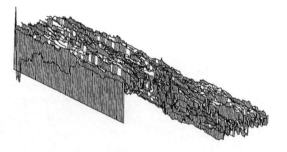

Fig A4.10 *'Mesh' of ĉ for case (ii) using Wagner's algorithm.*

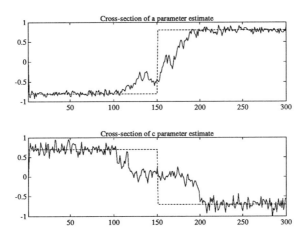

Fig A4.11 *Vertical cross-section of $\hat{a}$ and $\hat{c}$ for case (ii) using Wagner's algorithm.*

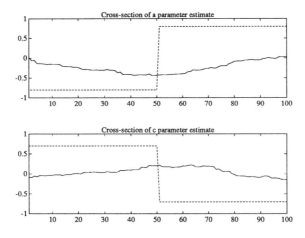

Fig A4.12 *Horizontal cross-section of $\hat{a}$ and $\hat{c}$ for case (ii) using Wagner's algorithm.*

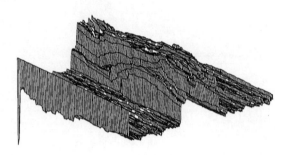

Fig A4.13 *'Mesh' of â for case (iv) using Wagner's algorithm.*

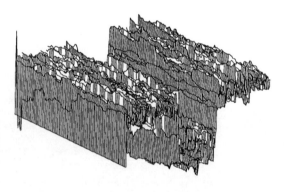

Fig A4.14 *'Mesh' of ĉ for case (iv) using Wagner's algorithm.*

Appendix 4 Simulation results for §4.6

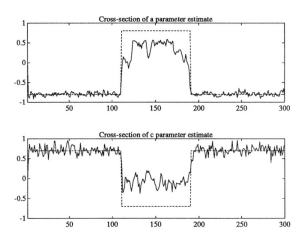

Fig A4.15 *Vertical cross-section of $\hat{a}$ and $\hat{c}$ for case (iv) using Wagner's algorithm.*

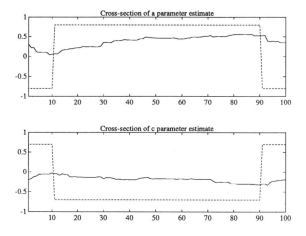

Fig A4.16 *Horizontal cross-section of $\hat{a}$ and $\hat{c}$ for case (iv) using Wagner's algorithm.*

212   Appendix 4 Simulation results for §4.6

Fig A4.17 *'Mesh' of $\hat{a}$ for case (i) using column forgetting.*

Fig A4.18 *'Mesh' of $\hat{c}$ for case (i) using column forgetting.*

Appendix 4 Simulation results for §4.6

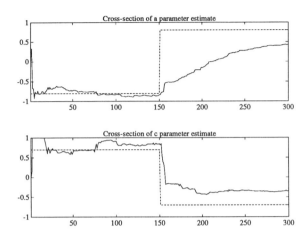

Fig A4.19 *Vertical cross-section of $\hat{a}$ and $\hat{c}$ for case (i) using column forgetting.*

Fig A4.20 *'Mesh' of $\hat{a}$ for case (ii) using column forgetting.*

Fig A4.21 *'Mesh' of $\hat{c}$ for case (ii) using column forgetting.*

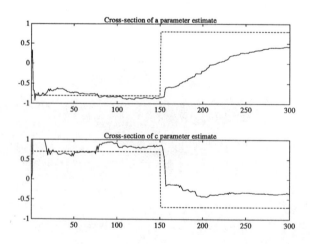

Fig A4.22 *Vertical cross-section of $\hat{a}$ and $\hat{c}$ for case (ii) using column forgetting.*

Appendix 4 Simulation results for §4.6

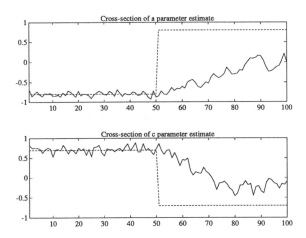

Fig A4.23 *Horizontal cross-section of â and ĉ for case (ii) using column forgetting.*

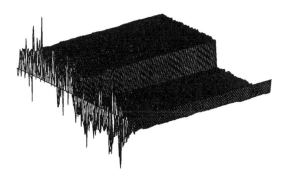

Fig A4.24 *'Mesh' of â for case (iii) using column forgetting.*

Fig A4.25 *'Mesh' of ĉ for case (iii) using column forgetting.*

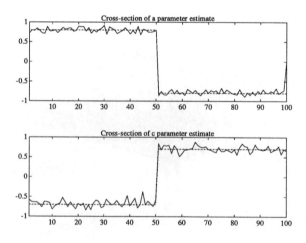

Fig A4.26 *Horizontal cross-section of â and ĉ for case (iii) using column forgetting.*

Appendix 4 Simulation results for §4.6

Fig A4.27 'Mesh' of $\hat{a}$ for case (iv) using column forgetting.

Fig A4.28 'Mesh' of $\hat{c}$ for case (iv) using column forgetting.

218   Appendix 4 Simulation results for §4.6

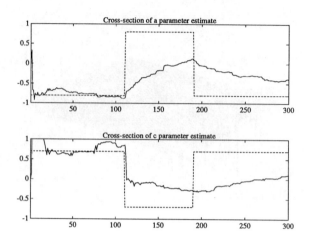

Fig A4.29 *Vertical cross-section of $\hat{a}$ and $\hat{c}$ for case (iv) using column forgetting.*

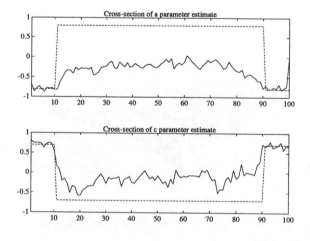

Fig A4.30 *Horizontal cross-section of $\hat{a}$ and $\hat{c}$ for case (iv) using column forgetting.*

Appendix 4 Simulation results for §4.6   219

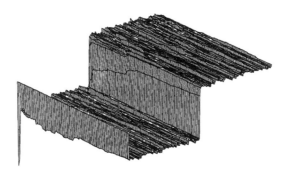

Fig A4.31 'Mesh' of $\hat{a}$ for case (i) using row forgetting.

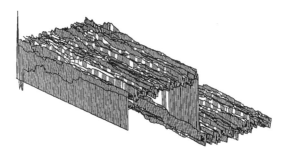

Fig A4.32 'Mesh' of $\hat{c}$ for case (i) using row forgetting.

Appendix 4 Simulation results for §4.6

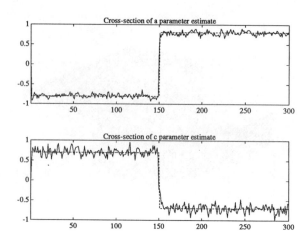

Fig A4.33 *Vertical cross-section of $\hat{a}$ and $\hat{c}$ for case (i) using row forgetting.*

Fig A4.34 *'Mesh' of $\hat{a}$ for case (ii) using row forgetting.*

Appendix 4 Simulation results for §4.6   221

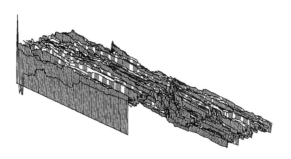

Fig A4.35 *'Mesh' of ĉ for case (ii) using row forgetting.*

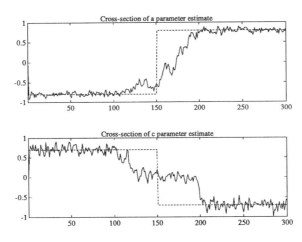

Fig A4.36 *Vertical cross-section of â and ĉ for case (ii) using row forgetting.*

222   Appendix 4 Simulation results for §4.6

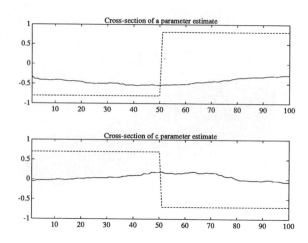

Fig A4.37 *Horizontal cross-section of $\hat{a}$ and $\hat{c}$ for case (ii) using row forgetting.*

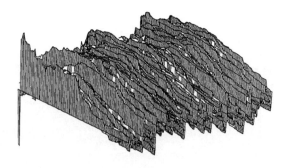

Fig A4.38 *'Mesh' of $\hat{a}$ for case (iii) using row forgetting.*

Appendix 4 Simulation results for §4.6

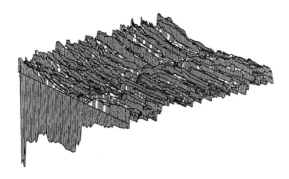

Fig A4.39 *'Mesh' of $\hat{c}$ for case (iii) using row forgetting.*

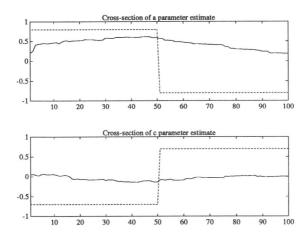

Fig A4.40 *Horizontal cross-section of $\hat{a}$ and $\hat{c}$ for case (iii) using row forgetting.*

224 Appendix 4 Simulation results for §4.6

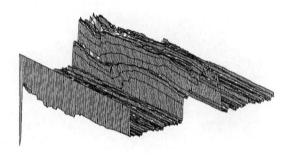

Fig A4.41 *'Mesh' of â for case (iv) using row forgetting.*

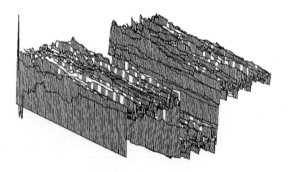

Fig A4.42 *'Mesh' of ĉ for case (iv) using row forgetting.*

Appendix 4 Simulation results for §4.6

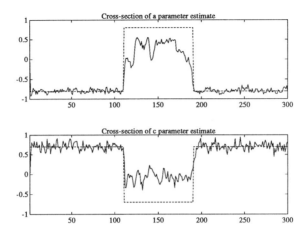

Fig A4.43 *Vertical cross-section of â and ĉ for case (iv) using row forgetting.*

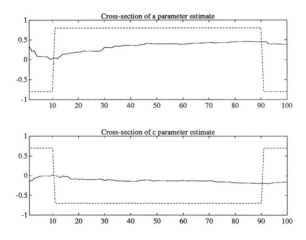

Fig A4.44 *Horizontal cross-section of â and ĉ for case (iv) using row forgetting.*

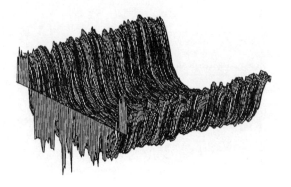

Fig A4.45 *'Mesh'* of $\hat{a}$ for case (v) using general two-dimensional forgetting.

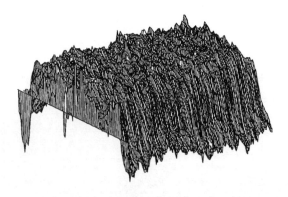

Fig A4.46 *'Mesh'* of $\hat{c}$ for case (v) using general two-dimensional forgetting.

Appendix 4 Simulation results for §4.6   227

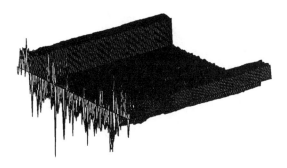

Fig A4.47 *'Mesh' of $\hat{a}$ for case (v) using column forgetting.*

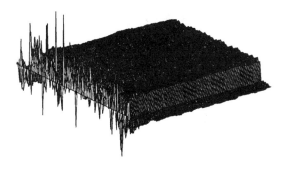

Fig A4.48 *'Mesh' of $\hat{c}$ for case (v) using column forgetting.*

228 Appendix 4 Simulation results for §4.6

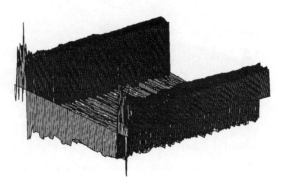

Fig A4.49 *'Mesh' of $\hat{a}$ for case (v) using the hybrid algorithm.*

Fig A4.50 *'Mesh' of $\hat{c}$ for case (v) using the hybrid algorithm.*

References

Angwin, D., 1989. Adaptive Image Restoration Using Reduced Order Model Based Kalman Filters. PhD Thesis, ECSE Dept, Rensselaer Polytechnic Institute, Troy NY, USA.

Ansell, H. G., 1964. On Certain Two-Variable Generalizations of Circuit Theory, with Applications to Networks of Transmission Lines and Lumped Reactances. IEEE Trans CT-11, pp214-223.

Astrom, K. J., 1970. Introduction to Stochastic Control Theory. Academic Press.

Astrom. K. J., Borisson, U., Ljung, L. and Wittenmark, B., 1977. Theory and Application of Self-Tuning Regulators. Automatica 13, pp457-476.

Astrom, K. J. and Wittenmark, B. 1973. On Self-Tuning Regulators. Automatica 9, pp185-199.

Attasi, S., 1975. Modelisation et Traitement des Suites a Deux Indices. IRIA Rapport Laboria, (Sept.).

Attasi, S., 1973. Systemes Lineaires Homogenes a Deux Indices. IRIA Rapport Laboria No.31, (Sept.).

Bierman, G. J., 1977. Factorization Methods for Discrete Sequential Estimation. Academic Press.

Bisiacco, M. and Fornasini, E., 1991. Structure and Solutions of the LQ Optimal Control Problem for 2D Systems. Kybernetika 27, pp231-242.

Borisson, U., 1979. Self-Tuning Regulators for a Class of Multivariable Systems. Automatica 15, pp209-215.

Caldas-Pinto, J. R., 1983. Self-Tuning Filters for Two Dimensions. PhD Thesis, CSC.

Clarke, D. W. and Gawthrop, P. J., 1975. Self-Tuning Controller. Proc IEE 122, pp929-934.

Clarke, D. W., Hodgson, A. J. F. and Tuffs, P. S., 1983. Offset Problem and k-Incremental Predictors in Self-tuning Control. Proc IEE 130 Pt D, pp217-225.

Clarke, D. W., Mohtadi, C. and Tuffs, P. S., 1987. Generalized Predictive Control: Part 1 the Basic Algorithm. Automatica 23, pp137-148.

Edmunds, J. M., 1976. Digital Adaptive Pole-Shifting Regulators. PhD Thesis, CSC.

Edwards, J. B. and Owens, D. H., 1982. Analysis and control of multipass processes. Wiley, Research Studies Press.

Fadali, M. S. and Gnanadekaran, R., 1989. Normal Matrices and their Stability Properties: Application to 2-D stabilization. IEEE Trans CAS-36, pp873-875.

Fonasini, E. and Marcesini, G. 1978. Doubly Indexed Dynamical Systems: State Space Models and Structural Properties. Math Systems Theory 12, pp59-75.

Fortescue, T. R., Kershenbaum, L. S. and Ydstie, B. E., 1981. Implementation of Self-Tuning Regulators with Variable Forgetting Factors. Automatica 17, pp831-835.

Fulton, W., 1969. Algebraic Curves. W A Benjamin, Inc.

Gawthrop, P. J., 1977. Some Interpretations of the Self-Tuning Controller. Proc IEE 124, pp889-894.

Gentleman, W. M., 1973. Least Squares Computations by Givens Transformations Without Square Roots. J. Inst. Maths Applics 12, pp329-336.

Golub, G. H. and van Loan, C. F., 1989. Matrix Computations (Second Edition). The Johns Hopkins University Press.

Goodman, D., 1977. Some Stability Properties of Two-Dimensional Linear Shift-Invariant Digital Filters. IEEE Trans CAS-24, pp201-208.

Hagglund, T., 1983. New Estimation Techniques for Adaptive Control. PhD Thesis, Dept of Automatic Control, Lund University, Sweden.

Heath, W. P., 1992. Self-Tuning Control for Two-Dimensional Processes. PhD Thesis, CSC.

Heath, W. P., 1993. Orthogonal Functions for Cross-Directional Control of Paper Making Machines and Plastic Film Extruders. CSC Report No. 787.

Holik, H., Weisshuhn, E. and Dahl, H., 1987. Influence of Headboxflow Conditions on Paper Properties and their Constancy. TAPPI Eng Conf, New Orleans.

Hwang, S. Y., 1981. Computation of Correlation Sequences in Two-Dimensional Digital Filters. IEEE Trans ASSP-29, pp58-61.

Kaczorek, T., 1985. Two-Dimensional Linear Systems. Lecture Notes in Control and Information Sciences, Springer-Verlag.

Kaczorek, T. and Klamka, J., 1986. Minimum Energy Control of 2-D Linear Systems with Variable Coefficients. IJC 44, pp645-650.

Kalman, R. E., 1958. Design of a Self-Optimizing Control System. Trans ASME 80, pp468-478.

Kastanakis, G. and Lizr, A., 1991. Interaction between the MD and CD Control Processes in Papermaking and Plastics Machines. Tappi Journal, pp77-83, (Feb.).

Koivo, H. N., 1980. A Multivariable Self-Tuning Controller. Automatica 16, pp351-366.

Kulhavy, R. and Karny, M., 1984. Tracking of Slowly Varying Parameters by Directional Forgetting. Proc 9th IFAC World Congress, pp78-83, Budapest, Hungary.

Kulhavy, R., 1987. Restricted Exponential Forgetting in Real-Time Identification. Automatica 23, pp589-600.

Kung, S.-Y., Levy, B. C., Morf, M. and Kailath, T., 1977. New Results in 2-D Systems Theory, Part II: 2-D State-Space Models—Realization and the Notions of Controllability, Observability and Minimality. Proc IEEE 65, pp945-961.

Lawson, C. L. and Hanson, R. J., 1974. Solving Least Squares Problems. Prentice-Hall.

Lee, E. B. and Lu, W.-S., 1985. Stabilisation of Two-Dimensional Systems. IEEE Trans AC-30, pp409-411.

Lee, E. B. and You, Y., 1988. Optimal Control of Two-Dimensional Systems. Proc 26th CDC, Los Angeles.

Lelic, M. A. and Zarrop, M. B., 1987. A Generalized Pole-Placement Self-Tuning Controller: Part 1 Basic Algorithm. IJC 46, pp569-607.

Levy, B. C., 1981. 2-D Polynomial and Rational Matrices, and their Applications for the Modelling of 2-D Dynamical Systems. PhD Thesis, Dept of Elec Engg, Stanford Univ, CA.

Li, C. and Fadali, M. S., 1991. Optimal Control of 2-D Systems. IEEE Trans AC-36, pp223-228.

Ljung, L. and Soderstrom, T., 1983. Theory and Practice of Recursive Identification. The MIT Press.

Lu, W.-S. and Lee, E. B., 1985. Stability Analysis for Two-Dimensional Systems in a Lyapunov Approach. IEEE Trans CAS-32, pp61-68.

Morf, M., Levy, B. C. and Kung, S.-Y., 1977. New Results in 2-D Systems Theory, Part I: 2-D Polynomial Matrices, Factorization and Coprimeness. Proc IEEE 65, pp861-872.

O'Connor, B. T. and Huang, T. S., 1981. Stability of General Two-Dimensional Recursive Filters. In Huang, T. S., (ed): Two-Dimensional Digital Signal Processing.1. Linear Filters. Springer-Verlag.

Ozaki, H. and Kadami, T., 1960. Positive Real Functions of Several Variables and their Applications to Variable Networks. IRE Trans CT-7, pp251-260.

Paraskevopoulos, P. N. and Kosimdou, O. I., 1981. Eigenvalue Assignment of Two-Dimensional Systems Using PID Controllers. Int J Syst Sci 12, pp407-422.

Peterka, V., 1970. Adaptive Digital Regulation of Noisy Systems. Proc 2nd IFAC Symp on Identification and Process Parameter Estimation, Prague.

Peterka, V., 1984. Predictor Based Self-Tuning Control. Automatica 20, pp39-50.

Rice, J. R., 1964. The Approximation of Functions. Addison-Wesley.

Richtmyer, R. D., 1957. Difference Methods for Initial-Value Problems. Interscience Publishers.

Roesser, R. P., 1975. A Discrete State-Space Model for Linear Image Processing. IEEE Trans AC-20, pp1-10.

Rogers, E. and Owens, D. H., 1989. 2D Transfer-Functions and Stability Tests for Differential Non-Unit Memory Multipass Processes. IJC 50, pp651-666.

Sebek, M., 1983. 2-D Polynomial Equations. Kybernetika 19, pp212-224.

Sebek, M., 1985. On 2-D Pole placement. Trans IEEE AC-30, pp819-822.

Smook, G. A., 1982. Handbook for Pulp and Paper Technologists. Tappi.

Soderstrom, T. and Stoica, P., 1989. System Identification. Prentice-Hall.

Tuffs, P. S. and Clarke, D. W., 1985. Self-tuning Control of Offset: a Unified Approach. Proc IEE 132 Pt D, pp100-110.

van der Waerden, B. L., 1964. Modern Algebra, 4th ed. (2 volumes). Frederic Ungar Publishing Co.

Wagner, G. R. and Wellstead, P. E., 1990. Extended Self-Tuning Predictors, Smoothers and Filters for Two-Dimensional Data Fields. International Journal of Adaptive Control and Signal Processing 4, pp299-329.

Wagner, G. R., 1987. Self-Tuning Algorithms for Two-Dimensional Signal Processing. PhD Thesis, CSC.

Wellstead, P. E., Prager, D. L. and Zanker, P. M., 1979. A Pole-Assignment Self-Tuning Regulator. Proc IEE 126, pp781-787.

Wellstead, P. E. and Sanoff, S. P., 1981. Extended Self-Tuning Algorithm. IJC 34, pp433-455.

Wellstead, P. E. and Caldas-Pinto, J. R. 1985. Self-Tuning Filters and Predictors for Two-Dimensional Systems, Part 1 Algorithms. IJC 42, pp457-478.

Wellstead, P. E. and Zarrop, M. B., 1991. Self-Tuning Systems. Wiley.

Westermeyer, W. N., 1987. Modellbildung und Simulation Der Papiermaschine zur Regelung des Flaechenbezogenen Massequerprofils der Papierbahn. PhD Thesis, Der Technischen Fakultaet der Universitaet Erlangen—Nurnberg.

Whittle, P., 1954. On Stationary Processes in the Plane. Biometrika 41, pp434-449.

Whittle, P., 1983. Prediction and Regulation by Linear Least-Square Methods. 2nd ed. University of Minnesota Press.

Wilhelm, R. G. and Fjeld, M., 1983. Control Algorithms for Cross Directional Control: The State of the Art. Proc 5th IFAC Conf. on Instrumentation and Automation in the Paper, Rubber, Plastics and Polymerization Industries, pp139-150.

Wilkinson, J. H., 1977. Some Recent Advances in Numerical Linear Algebra. In Jacobs, D. A. H., (ed): The State of the Art in Numerical Analysis, pp1-53. Academic Press.

Wittenmark, B., 1974. A Self-Tuning Predictor. IEEE Trans AC-19, pp848-851.

Index

Admissible controller, 11, 75, 77
Algebra of two-dimensional systems, 6, 13, 17–19, 26–27, 37, 80, 82, 170, 172
AML (approximate maximum likelihood), 83–84, 87, 95, 113, 171–172
AR (auto-regressive) model, 83–84, 89, 113
ARMA (auto-regressive moving average) model, 1, 12, 15–17, 61, 83, 87, 89, 113
Attasi's two-dimensional state-space model, 94, 96–99

Backwards shift operator, 10
Basis weight, 5–6, 158
Bierman U/D factorization, 102

CARMA (controlled auto-regressive moving average) model, 3, 9, 12, 32–33, 61, 137, 170
Chebyshev polynomials, 166
Cloth weaving, 4
Coating processes, 4
Column forgetting, 91, 110, 112–117, 125–127, 129, 171
Column weighting, 152–154, 164–165

Convergence properties, 172
Coprimeness, 13–14
Covariance matrix, 102, 110, 137–139, 144, 165–167

Data
    with finite edges, 7, 32–82, 86, 113–116, 125–128, 146, 148–151, 161–169, 171–172
    with no edges, 6, 9–31, 37–45, 54–60
Delay, 10, 72, 77
Delay-differential systems, 1
Diophantine equation, 9, 17–21, 24, 26–27, 170
Directional forgetting, 89

Final value theorem, 150
Finite-difference approximation methods, 64, 72
Forgetting strategies, 7–8, 83, 88–135
Fundamental theorem of algebra, 13

General two-dimensional forgetting, 91–96, 98–109, 116–117, 119–121, 125–128, 130–131, 134–135, 171
Geophysical data, 1
Givens rotations, 107–109

Global support, 10
GMV (generalised minimum variance)
  control, 3, 7–9, 30–32, 37,
  39–41, 43, 45, 51–53, 56, 70–82,
  136–153, 155–169, 171–172
  explicit, 137–147
  implicit, 144–147

Householder transformation, 104–107
Hybrid algorithm, 125–127

Image processing, 1, 3, 8, 171
Incremental model, 158–164
Information matrix, 83, 95–96, 102,
  138, 171
Integral control, 136, 151, 153, 155–158,
  160–162, 167, 172

Lambda controller, 43
Legendre polynomials, 166–167, 169
Local support, 10, 61–68, 70–82,
  148–149

MA (moving average) model, 82
Matrix inversion lemma, 85, 92, 102,
  110, 112
Minimum variance control, 3, 7–9,
  28–30, 32, 34–37, 41–44, 50–51,
  56, 136, 170
Multipass processes, 2, 8
Multivariable model, 8, 32, 43, 46–60,
  80, 151–153, 163–164, 171

NSHP (non-symmetric half-plane),
  9–12, 15–17, 27, 48–49, 61–68,
  70–82, 170, 172

Offset handling, 7–8, 136, 158–169, 171

Papermaking, 4–6, 8, 10, 65, 68–69, 77,
  158, 172
Parameter estimation, 7, 57, 83–136,
  139, 141–146, 148–149, 159,
  162–171
Partial differential equations, 1, 64, 152
Persistent excitation, 139, 144
Plastic-film extrusion, 4
Pole-assignment control, 2–3, 6, 8–9,
  19–21
Prediction, 3, 6–9, 21–28, 32, 34–39,
  47–49, 54–56, 60, 136, 170–171
Predictive control, 3
Pseudo-output, 30, 39–40, 53, 70, 74,
  77, 80, 82, 138, 145, 159, 161,
  164

QR factorisation, 83, 102–109, 171
Quarter-plane, 11–17, 27

Raster scan, 11, 88, 170
RELS (recursive extended least
  squares), 83, 87, 95, 171
RLS (recursive least squares), 83, 86–87
Roesser's two-dimensional state-space
  model, 96–102
Row forgetting, 91, 110–111, 116–117,
  126–127, 171

Self-tuning control, 3, 6–8, 136–153,
  155–171
Setpoint tracking, 7–8, 136, 149–153,
  155–158, 162, 171

Index

SHP (symmetric half-plane), 61–66, 68, 70–82
Spatial non-causality, 32, 61–82
Stability, 13–15, 28–29, 31, 37, 43, 58–59, 156–158, 160, 172
Static model, 158–159, 162–169
Steel rolling, 4
SWP (symmetric wedged-plane), 62–65, 67–68, 70–82, 113

Transform from quarter-plane to NSHP causal process, 7, 12, 15–17

Variance calculation, 7, 12–13, 15–17, 26, 41, 43

Wagner's algorithm, 7, 83, 88–89, 116–117, 122–127, 132–133, 171
Web manufacture, 4, 8
Weighted least squares, 7–8, 89